Design Recipes for FPGAs
Using Verilog and VHDL

Design Recipes for FPGAs

Using Verilog and VHDL

Peter Wilson

ELSEVIER

AMSTERDAM • BOSTON • HEIDELBERG • LONDON
NEW YORK • OXFORD • PARIS • SAN DIEGO
SAN FRANCISCO • SINGAPORE • SYDNEY • TOKYO
Newnes is an imprint of Elsevier

Newnes

Newnes is an imprint of Elsevier
32 Jamestown Road, London NW1 7BY, UK
525 B Street, Suite 1800, San Diego, CA 92101-4495, USA
225 Wyman Street, Waltham, MA 02451, USA
The Boulevard, Langford Lane, Kidlington, Oxford OX5 1GB, UK

First edition 2007
Second edition 2016

Library of Congress Cataloging-in-Publication Data
A catalog record for this book is available from the Library of Congress

British Library Cataloguing in Publication Data
A catalogue record for this book is available from the British Library

For information on all Newnes publications
visit our website at http://store.elsevier.com/

Printed and bound in the United Kingdom

ISBN: 978-0-08-097129-2

Publisher: Todd Green
Acquisitions Editor: Tim Pitts
Editorial Project Manager: Charlotte Kent
Production Project Manager: Lisa Jones
Designer: Mark Rogers

Contents

PART 5 FUNDAMENTAL TECHNIQUES   283

Preface to the Second Edition

The original idea behind the first edition of this book was to collect some of the useful methods for designing digital systems using FPGAs that I had accumulated over the years and had been passing on to students in our courses at the University of Southampton. As a result, the original book was written using VHDL, as this was very often the hardware description language of choice for university students and for many courses (as was the case at Southampton).

The intervening time has seen the development of other options, such as System-C or System-Verilog (plus the continuing popularity of Verilog). One of the common questions to me was "Why is there not a Verilog edition of this book?". I have therefore taken the opportunity with the second edition to introduce Verilog, to widen the applicability of the book to as many designers as possible.

The second edition also offers the chance to correct errors and take on board the numerous reviews over the past seven years since the first edition was published. For these comments and suggestions I am most grateful to the readers of the book. FPGAs have also moved on in leaps and bounds since the first edition, and this also gives an opportunity to update some of the technological background and correct errors in the first edition.

Above all else, this book was not and is not intended to be a textbook for digital systems design, but rather a useful handbook for designers to dip in and use wherever it can help.

I sincerely hope you find this book useful and good luck with your FPGA designs!

Peter Wilson
University of Bath

Preface to the First Edition

This book is designed to be a desktop reference for engineers, students and researchers who use field programmable gate arrays (FPGAs) as their hardware platform of choice. This book has been produced in the spirit of the "numerical recipe" series of books for various programming languages – where the intention is not to teach the language *per se*, but rather the philosophy and techniques required in making your application work. The rationale of this book is similar in that the intention is to provide the methods and understanding to enable the reader to develop practical, operational VHDL that will run correctly on FPGAs.

It is important to stress that this book is *not* designed as a language reference manual for VHDL. There are plenty of those available and I have referenced them throughout the text. This book is intended as a reference for design *with* VHDL and can be seen as complementary to a conventional VHDL textbook.

Acknowledgments

The first acknowledgement I must make is to the colleagues and students who have inspired, supported, and tolerated my teaching over the past 15 years at the Universities of Southampton and Bath. In no particular order, some of the key people in that process have been Professor Andrew Brown, Dr. Reuben Wilcock, Dr. Neil Ross, Professor Alan Mantooth, and Professor Mark Zwolinski. I am truly grateful.

I must also note the kind and tolerant Elsevier staff who have cajoled me into this second edition, including Tim Pitts, Charlotte Kent, and Lisa Jones, plus all the reviewers who provided helpful comments during this process—thanks.

Finally, a big "thank you" goes to my wife Caroline and daughter Heather. Without their support and love, none of this would be possible.

Peter Wilson
University of Bath

PART 1

Overview

The first part of the book provides a starting point for engineers who may have some digital experience but not necessarily with FPGAs in particular, or with either of the languages featured in this book (VHDL and Verilog). While the book is not intended to teach either language, "primers" are given in both as *aides de memoire* to get started. An overview of the main design approaches and tool flows is also provided as a starting point.

Introduction

1.1 Overview

The book is divided into five main parts. In the introductory part of the book, primers are given on FPGAs (field-programmable gate arrays), Verilog and the standard design flow. In the second part of the book, a series of complex applications that encompass many of the key design problems facing designers today are worked through from start to finish in a practical way. This will show how the designer can interpret a specification, develop a top-down design methodology and eventually build in detailed design blocks perhaps developed previously or by a third party. In the third part of the book, important techniques are discussed, worked through and explained from an example perspective so you can see exactly how to implement a particular function. This part is really a toolbox of advanced specific functions that are commonly required in modern digital design. The fourth part on advanced techniques discusses the important aspect of design optimization, that is, how can I make my design faster, or more compact? The fifth part investigates the details of fundamental issues that are implemented in VHDL and Verilog. This final part is aimed at designers with a limited VHDL or Verilog coding background, perhaps those looking for simpler examples to get started, or to solve a particular detailed issue.

1.2 Verilog vs. VHDL

One of the longest standing "arguments" between engineers in digital design has been the issue of which is best—Verilog or VHDL? For many years this was partly a geographical divide, with North America seeming to be mainly using Verilog and Europe going more for VHDL, although this was not universal by any means. In many cases, the European academic community was trending toward VHDL with its easy applicability to system level design, and the perception that Verilog was really more a "low level" design language. With the advent of SystemVerilog and the proliferation of design tools, these boundaries and arguments have largely subsided, and most engineers realize that they can use IP blocks from either language in most of the design tools. Of course, individuals will always have their own preferences; however it is true to say that now it is genuinely possible to be language agnostic and use whichever language and tools the user prefers. More often than not, the choice will depend on

Design Recipes for FPGAs. http://dx.doi.org/10.1016/B978-0-08-097129-2.00001-5

three main factors: (a) the experience of the user (for example, they may have a background in a particular language); (b) the tools available (for example, some tool flows may simply work better with a particular language—SystemVerilog for instance may not be supported by the tools available); and (c) corporate decisions (where the company or institution has a preference for a specific language, and in turn this may mean that libraries must be in a specific format and language). For researchers, there is a plethora of information on all design languages available, with many example designs published on the web, making it relatively simple to use one or another of the main languages, and sometimes even a mixture of languages (using precompiled libraries, for example). Of course, this is also available to employees of companies and free material is now widely available from sources such as Open Cores (http://www.opencores.org), the Free Model Foundry (http://www.freemodelfoundry. com/) and the Open Hardware Repository at CERN (http://www.ohwr.org/).

1.3 Why FPGAs?

There are numerous options for designers in selecting a hardware platform for custom electronics designs, ranging from embedded processors, application specific integrated circuits (ASICs), programmable microprocessors (PICs), FPGAs to programmable logic devices (PLDs). The decision to choose a specific technology such as an FPGA should depend primarily on the design requirements rather than a personal preference for one technique over another. For example, if the design requires a programmable device with many design changes, and algorithms using complex operations such as multiplications and looping, then it may make more sense to use a dedicated signal processor device such as a DSP that can be programmed and reprogrammed easily using C or some other high level language. If the speed requirements are not particularly stringent and a compact cheap platform is required, then a general purpose microprocessor such as a PIC, AVR, or MBED would be an ideal choice. Finally, if the hardware requirements require a higher level of performance, say up to several hundred megahertz operation, then an FPGA offers a suitable level of performance, while still retaining the flexibility and reusability of programmable logic.

Other issues to consider are the level of optimization in the hardware design required. For example, a simple software program can be written in C and then a microprocessor programmed, but the performance may be limited by the inability of the processor to offer parallel operation of key functions. This can be implemented much more directly in an FPGA using parallelism and pipelining to achieve much greater throughput than would be possible using a microprocessor. A general rule of thumb when choosing a hardware platform is to identify both the design requirements and the possible hardware options and then select a suitable platform based on those considerations. For example, if the design requires a basic clock speed of up to 1 GHz then an FPGA would be a suitable platform. If the clock speed could be 3-4 MHz, then the FPGA may be an expensive (overkill) option. If the design requires a flexible processor option, although the FPGAs available today support embedded processors, it probably makes sense to use a DSP or microprocessor. If the design requires dedicated hardware functionality, then an FPGA is the route to take.

If the design requires specific hardware functions such as multiplication and addition, then a DSP may well be the best route, but if custom hardware design is required, then an FPGA would be the appropriate choice. If the design requires small simple hardware blocks, then a PLD or CPLD may be the best option (compact, simple programmable logic); however, if the design has multiple functions, or a combination of complex controller and specific hardware functions, then the FPGA is the route to take. Examples of this kind of decision can be dependent on the complexity of the hardware involved. For example, a high performance signal processor with multiple parallel tasks will probably require an FPGA rather than a PLD device, simply due to the complexity of the hardware involved. Another related issue is that of flexibility and programmability. If an FPGA is used, and the resources are not used up on a specific device (say up to 60% for example), if a communications protocol changes, or is updated, then the device may well have enough headroom to support additional features, or updates, in the future.

Finally, the cost of manufacture will be important for products in the field, as well as where the device is deployed (in terms of the overall weight, power requirements, footprint, and volume). Also, the need for upgrading firmware may mandate an FPGA to allow this to be done easily. The use of an FPGA also allows much higher performance, particularly on high speed links or memory, enabling the design to be somewhat tolerant of future changes.

1.4 Summary

Using the simple guidelines and with the aid of some of the examples in this book, an engineer can hopefully make an intelligent choice about the best platform to choose, and also which hardware device to select based on these assumptions. A nice aspect of most FPGA design software packages is that multiple design platforms can be evaluated for performance and utilization prior to making a final decision on the hardware of choice. This book will show how both VHDL and Verilog can be used to solve typical design problems, and hopefully will help designers get their own designs completed faster and more efficiently.

CHAPTER 2

An FPGA Primer

2.1 Introduction

This section is an introduction to the Field Programmable Gate Array (FPGA) platform for those unfamiliar with the technology. It is useful when designing hardware to understand that the context that the hardware description language models (VHDL or Verilog) is important and relevant to the ultimate design.

2.2 FPGA Evolution

Since the inception of digital logic hardware in the 1970s, there has been a plethora of individual semiconductor digital devices leading to the ubiquitous TTL logic series still in use today (74/54 series logic), now extended to CMOS technology (HC, AC, FC, FCT, HCT, and so on). While these have been used extensively in printed circuit board (PCB) design and still are today, there has been a consistent effort over the last 20 years to introduce greater programmability into basic digital devices.

One of the reasons for this need is the dichotomy resulting from the two differing design approaches used for many digital systems. On the hardware side, the drive is usually toward ultimate performance, that is, faster, smaller, lower power, and cheaper. This often leads to custom integrated circuit design (Application Specific Integrated Circuits or ASICs) where each chip (ASIC) has to be designed, laid out, fabricated, and packaged individually. For large production runs this is very cost effective, but obviously this approach is hugely expensive (masks alone for a current silicon process may cost over $500,000) and time consuming (can take up to a year or even more for large and complex designs).

From a software perspective, however, a more standard approach is to use a standard processor architecture such as Intel Pentium, PowerPC or ARM, and develop software applications that can be downloaded onto such a platform using standard software development tools and cross compilers. This type of approach is obviously quicker to implement an initial working platform; however, usually there is a significant overhead due to the need for operating systems, compiler inefficiency and also a performance reduction due to the indirect relationship between the hardware and the software on the processor. The other

Design Recipes for FPGAs. http://dx.doi.org/10.1016/B978-0-08-097129-2.00002-7

issue from a hardware perspective is often the compromise necessary when using a standard platform, for example will it be fast enough? Another key issue when designing hardware is having access to that hardware. In many processor platforms, the detailed hardware is often difficult to access directly or efficiently enough to meet the performance needs of the system, and with the rigid architecture in terms of data bus and address bus widths on standard processors, very often there is no scope for general purpose IO (Inputs and Outputs) which are useful for digital designers.

As a result, programmable devices have been developed as a form of intermediate approach: hardware design on a high-performance platform, optimal resources with no operating system required and reconfigurable as the devices can be reprogrammed.

2.3 Programmable Logic Devices

The first type of device to be programmable was the Programmable Array Logic (PAL) with a typical layout as shown in Figure 2.1. This consists of an array of logic gates that could be connected using an array of connections. These devices could support a small number of flip-flops (usually <10) and were able to implement small state machines. These devices still have a use for specific functions on a small scale, but clearly will be limited for more complex applications. They are, however, still useful for low-cost and compact solutions to a specific digital design requirement.

Complex Programmable Logic Devices (CPLD) such as shown in Figure 2.2 were developed to address the limitations of simple PAL devices. These devices used the same basic principle as PALs, but had a series of macro blocks (each roughly equivalent to a PAL) that were connected using routing blocks. With, in some cases, many thousands of logical elements, the CPLD can be extremely useful for implementing a programmable device with custom logic functions and state machines. In some ways, the latest CPLD and early FPGA devices are almost indistinguishable, with one crucial difference. The CPLD is a fixed array of logic, but the FPGA uses complex logic blocks (discussed in the next section of this chapter). However, CPLDs are still of a relatively small scale, and the modern reconfigurable device of choice for high performance is the FPGA.

2.4 Field Programmable Gate Arrays

Field Programmable Gate Arrays (FPGAs) were the next step from CPLDs. Instead of a fixed array of gates, the FPGA uses the concept of a Complex Logic Block (CLB). This is configurable and allows not only routing on the device, but also each logic block can be configured optimally. A typical CLB is shown in Figure 2.3. This extreme flexibility is very efficient as the device does not rely on the fixed logical resources (as in the case of a CPLD),

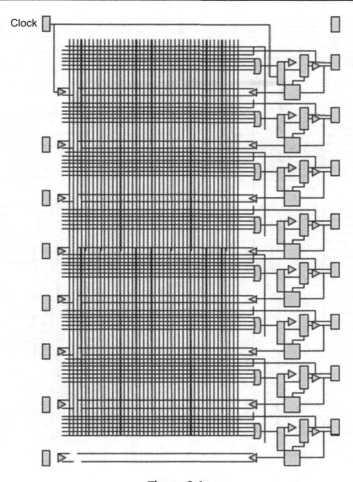

Figure 2.1
Typical programmable logic device.

but rather is able to define whichever logical functions are required as part of the logic block reconfiguration.

The CLB has a look-up table (LUT) that can be configured to give a specific type of logic function when programmed. There is also a clocked d-type flip flop that allows the CLB to be combinatorial (non-clocked) or synchronous (clocked), and there is also an enable signal. A typical commercial CLB (in this case from Xilinx®) is shown in Figure 2.4 and this shows clearly the two 4 input LUTs and various multiplexers and flip flops in a real device.

A typical FPGA will have hundreds or thousands of CLBs, of different types, on a single device allowing very complex devices to be implemented on a single chip and configured

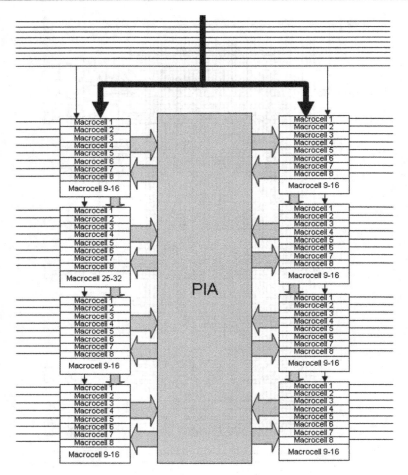

Figure 2.2
Complex programmable logic device.

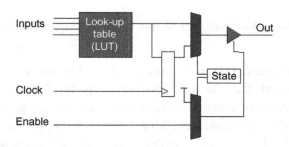

Figure 2.3
FPGA complex logic block.

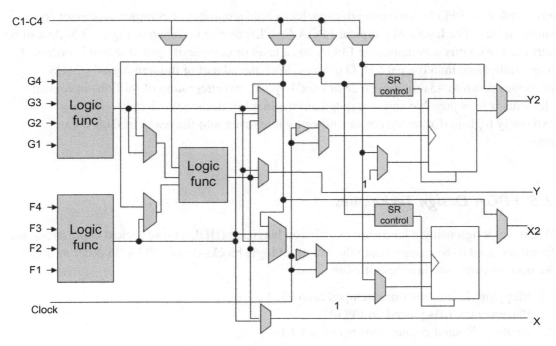

Figure 2.4
Typical commercial CLB architecture.

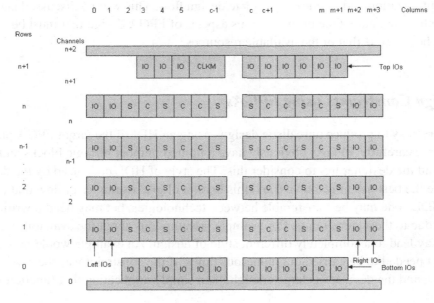

Figure 2.5
FPGA structure of CLB.

easily. Modern FPGAs have enough capacity to hold a number of complex processors on a single device. The layout of a typical FPGA (in CLB terms) is shown in Figure 2.5. As can be surmised from this schematic, the FPGA has a level of complexity possible that is orders of magnitude more than typical CPLD devices. With the advent of modern digital CMOS processes down to 45 nm or even 28 nm and beyond, the integration of millions of logical elements is now possible and to a speed unimaginable a decade previously—making extremely high-performance devices now possible (even into the realm of Gb/s data rates).

2.5 FPGA Design Techniques

When we design using a hardware description language (HDL), these logical expressions and functions need to be mapped onto the low level logic blocks on an FPGA. In order to do this, we need to carry out three specific functions:

1. Mapping: Logic functions mapped onto CLBs.
2. Placement: CLBs placed on FPGA.
3. Routing: Routed connections between CLBs.

It is clearly becoming impossible to design "by hand" using today's complex designs; we therefore rely on synthesis software to turn our HDL design description into the logic functions that can be mapped onto the FPGA CLBs. This design flow is an iterative process including optimization and implies a complete design flow. This will be discussed later in this book in much more detail. One of the obvious aspects of FPGA design that must be considered, however, is that of the available resources.

2.6 Design Constraints using FPGAs

It can be very easy to produce unrealistic designs using an HDL if the target FPGA platform is not considered carefully. FPGAs obviously have a limited number of logic blocks and routing resources, and the designer has to consider this. The style of HDL code used by the designer should make the best use of resources, and this book will give examples of how that can be achieved. HDL code may be transferable between technologies, but may need rewriting for best results due to these constraints. For example, assumptions about the availability of resources may lead to a completely different style of design. An example would be a complex function that needed to be carried out numerous times. If the constraint was the raw performance, and the device was large enough, then simply duplicating that function in the hardware would enable maximum data rates to be achieved. On the other hand, if the device is very small and can only support a smaller number of functions, then it would be up to the designer to consider pipelining or resource sharing to enable the device to be programmed, but obviously this would be at the cost of raw performance. The constraints placed on the designer by the FPGA platform itself can therefore be a significant issue in the choice of device or development platform.

2.7 Development Kits and Boards

There are now a wide array of development kits to suit all levels of budget and performance requirements from the manufacturers themselves, or from third-party companies specializing in development kits and board design. With the FPGA manufacturers being proactive in providing design software on the web (often for free for noncommercial purposes), it has become much less of a hurdle for engineers to obtain access to both the design tools and the hardware to test out their concepts.

One of the major advantages with the modern development boards is that they tend to have an FPGA device that can generally handle almost all the major building blocks (processors, display drivers, network stacks) even on a relatively low-end device. The beauty of the boards too is that with the development of multiple layer PCB designs, most of the common interface elements can also be integrated on a very small board. With both Xilinx and Altera supporting credit card sized boards, these are well within the reach of students and engineers on a very small budget. Mid-Range boards are also available for more lab based usage, such as the DE series of boards from Terasic, based around the Altera FPGA devices, starting with the credit card sized DE0-Nano, DE0 and continuing up in power and complexity. An excellent starter board, for example, is the DE0 board, which is slightly larger than its DE0 successor, but perhaps a little easier to use in terms of access to switches and plugs. This board has two 40-way GPIO (General Purpose Input Output) connectors for general interfacing, a VGA output, PS/2 input, Ethernet, USB, SD Card socket and a selection of LEDs, switches, and buttons. This board is shown in Figure 2.6. There is a series of boards available for the Xilinx FPGAs, with similar ranges of options with an example being the Nexsys 3™ board from Digilent® which has a similar range of IO capability to the Altera based boards. The Nexsys 3™ board is shown in Figure 2.7.

With such an extensive range of options and prices, it is now a matter of choice in many cases which platform to use. Each one will have its strengths and weaknesses, and the designer is able to select the device and board to develop their own design, taking into account their own requirements and constraints.

Figure 2.6
DE0 development board from Terasic (Altera FPGA device).

Figure 2.7
Nexsys 3 development board from Digilent (Xilinx FPGA device).

2.8 Summary

This chapter introduces the basic technology behind FPGAs and their development. The key design issues are highlighted and some of the important design techniques introduced. Later chapters in this book will develop these in more detail either from a detailed design perspective or from a methodology point of view.

CHAPTER 3

A VHDL Primer: The Essentials

3.1 Introduction

This chapter of the book is not intended as a comprehensive VHDL reference book as there are many excellent texts available that fit that purpose, including Zwolinski [1], Navabi [2], or Ashenden [3] (full details are provided in the References heading).

Instead, this chapter is designed to give concise and useful summary information on important language constructs and usage in VHDL, hopefully helpful and easy to use, but not necessarily comprehensive. The information is helpful enough for the reader to understand the key concepts in this book; however, I would thoroughly recommend obtaining access to a textbook on VHDL or Verilog if the reader is serious in becoming expert in HDL design for digital systems. This book is intended as a complement to a textbook.

This chapter will introduce the key concepts in VHDL and the important syntax required for most VHDL designs, particularly with reference to FPGAs. In most cases, the decision to use VHDL over other languages such as Verilog or SystemC will have less to do with designer choice and more to do with software availability and company decisions. Over the last decade or so, a war of words has raged between the VHDL and Verilog communities about which is the best language, and in most cases it is completely pointless as the issue is more about design than syntax. There are numerous differences in the details between VHDL and Verilog, but the fundamental philosophical difference historically has been the design context of the two languages.

Verilog has come from a bottom-up tradition and has been heavily used by the IC industry for cell-based design, whereas the VHDL language has been developed much more from a top-down perspective. Of course, these are generalizations and largely out of date in a modern context, but the result is clearly seen in the basic syntax and methods of the two languages. While this has possibly been the case in the past, with the advent of the higher level "SystemVerilog" variant of Verilog providing much of the same capability as VHDL at the system level, this has also become popular.

Unfortunately, while there are many languages now available to designers, most of the FPGA design tools support subsets, and therefore in some cases support for SystemVerilog may be

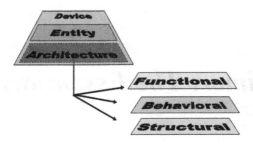

Figure 3.1
VHDL models with different architectures.

patchy. It is therefore useful to describe using VHDL and Verilog; however, this book will also provide some introductory material to SystemVerilog for completeness.

Without descending into a minute dissection of the differences between Verilog and VHDL, one important aspect of VHDL is the ability to use multiple levels of model with different architectures as shown in Figure 3.1.

This is not unique to VHDL, and in fact Verilog does have the concept of different behavior in a single module; however, it is explicitly defined in VHDL and is extremely useful in putting together practical multi-level designs in VHDL. The division of a model into its interface part (the *entity* in VHDL) and the behavior part (the *architecture* in VHDL) is an incredibly practical approach for modeling multiple behavior for a single interface and makes model exchange and multiple implementation practical.

The remainder of this chapter will describe the key parts of VHDL, starting with the definition of a basic model structure using entities and architectures, discuss the important variable types, review the methods of encapsulating concurrent, sequential and hierarchical behavior and finally introduce the important fundamental data types required in VHDL.

3.2 Entity: Model Interface

3.2.1 The Entity Definition

The VHDL entity defines how a design element described in VHDL connects to other VHDL models and also defines the name of the model. The entity also allows the definition of any parameters that are to be passed into the model using hierarchy. The basic template for an entity is as follows:

```
1    entity <name> is
2    ...
3    end entity <name>;
```

If the entity has the name *test* then the entity template could be either:

```
1    entity test is
2    end entity test;
```

or:

```
1    entity test is
2    end test;
```

3.2.2 Ports

The method of connecting entities together is using *ports*. These are defined in the entity using the following method:

```
1    port (
2    -- list of port declarations
3    );
```

The port declaration defines the type of connection and direction where appropriate. For example, the port declaration for an input bit called in1 would be:

```
1    in1 : in bit;
```

And if the model had two inputs (in1 and in2) of type bit and a single output (out1) of type bit then the declaration of the ports would be defined as follows:

```
1    port (
2            in1, in2 : in bit;
3            out1 : out bit
4    );
```

As the connection points between entities are effectively the same as those inter-process connections, they are effectively signals and can be used as such within the VHDL of the model.

3.2.3 Generics

If the model has a parameter, then this is defined using generics. The general declaration of generics is shown below:

```
1    generic (
2    -- list of generic declarations
3    );
```

In the case of generics, the declaration is similar to that of a constant with the form as shown below:

```
1    param1 : integer := 4;
```

Taking an example of a model that had two generics, gain (integer) and time_delay (time), they could be defined in the entity as follows:

```
1    generic (
2      gain : integer := 4;
3      time_delay : time = 10 ns;
4    );
```

3.2.4 Constants

It is also possible to include model specific constants in the entity using the standard declaration of constants method previously described; for example:

```
1    constant : rpullup : real := 1000.0;
```

3.2.5 Entity Examples

To illustrate a complete entity, we can bring together the ports and generics examples previously shown and construct the complete entity for this example:

```
1    entity test is
2      port (
3        in1, in2 : in bit;
4        out1 : out bit;
5      );
6      generic (
7        gain : integer := 4;
8        time_delay : time := 10 ns;
9      );
10     constant : rpullup : real := 1000.0;
11     end entity test;
```

3.3 Architecture: Model Behavior

3.3.1 Basic Definition of An Architecture

While the entity describes the interface and parameter aspects of the model, the architecture defines the behavior. There are several types of VHDL architecture and VHDL allows different architectures to be defined for the same entity. This is ideal for developing behavioral, RTL, and gate level architectures that can be incorporated into designs and tested using the same test benches.

The basic approach for declaring an architecture could be as follows:

```
1    architecture behavior of test is
2      -- architecture declarations
3    begin
4      -- architecture contents
5    end architecture behavior;
```

or

```
1    architecture behavior of test is
2      -- architecture declarations
3    begin
```

```
4       -- architecture contents
5   end behavior;
```

3.3.2 Architecture Declaration Section

After the declaration of the architecture name and before the begin statement, any local signals or variables can be declared. For example, if there were two internal signals to the architecture called sig1 and sig2, they could be declared in the declaration section of the model as follows:

```
1   architecture behavior of test is
2           signal sig1, sig2 : bit;
3   begin
```

Then the signals can be used in the architecture statement section.

3.3.3 Architecture Statement Section

VHDL architectures can have a variety of structures to achieve different types of functionality. Simple combinatorial expressions use signal assignments to set new signal values as shown below:

```
1   out1 <= in1 and in2 after 10 ns;
```

Note that for practical design, the use of the *after 10 ns* statement is not synthesizable. In practice, the only way to ensure correct synthesizable design is to either make the design delay insensitive or synchronous. The design of combinatorial VHDL will result in additional delays due to the technology library gate delays, potentially resulting in glitches or hazards. An example of a multiple gate combinatorial architecture, using internal signal declarations is given below:

```
1   architecture behavioral of test is
2       signal int1, int2 : bit;
3   begin
4       int1 <= in1 and in2;
5       int2 <= in3 or in4;
6       out1 <= int1 xor int2;
7   end architecture behavioral;
```

3.4 Process: Basic Functional Unit in VHDL

The process in VHDL is the mechanism by which sequential statements can be executed in the correct sequence, and with more than one process, concurrently. Each process consists of a sensitivity list, declarations, and statements. The basic process syntax is given below:

```
1   process sensitivity_list is
2       -- declaration part
```

```
3    begin
4      -- statement part
5    end process;
```

The sensitivity list allows a process to be activated when a specific signal changes value; for example a typical usage would be to have a global clock and reset signal to control the activity of the process, as follows:

```
1    process (clk, rst) is
2    begin
3      -- process statements
4    end process;
```

In this example, the process would only be activated when either clk or rst changed value. Another way of encapsulating the same behavior is to use a wait statement in the process so that the process is automatically activated once, and then waits for activity on either signal before running the process again. The same process could then be written as follows:

```
1    process
2    begin
3      -- process statements
4      wait on clk, rst;
5    end process;
```

In fact, the location of the wait statement is not important, as the VHDL simulation cycle executes each process once during initialization, and so the wait statement could be at the start or the end of the process and the behavior would be the same in both cases.

In the declaration section of the process, signals and variables can be defined locally as described previously; for example, a typical process may look like the following:

```
1    process (a) is
2      signal na : bit;
3    begin
4      na <= not a;
5    end process;
```

The local signal na and the process are activated by changes on the signal a which is externally declared (with respect to the process).

3.5 Basic Variable Types and Operators

3.5.1 Constants

When a value needs to be static throughout a simulation, the type of element to use is a constant. This is often used to initialize parameters or to set fixed register values for comparison. A constant can be declared for any defined type in VHDL with examples as follows:

```
1    constant a : integer := 1;
2    constant b : real := 0.123;
3    constant c : std_logic := 0;
```

3.5.2 Signals

Signals are the link between processes and sequential elements within processes. They are effectively wires in the design and connect all the design elements together. When simulating signals, the simulator will in turn look at updating the signal values and also checking the sensitivity lists in processes to see whether any changes have occurred that will mean that processes become active.

Signals can be assigned immediately or with a time delay, so that an event is scheduled for sometime in the future (after the specified delay). It is also important to recognize that signals are not the same as a set of sequential program code (such as in C), but are effectively concurrent signals that will not be able to be considered stable until the next time the process is activated.

Examples of signal declaration and assignment are shown below:

```
1    signal sig1 : integer := 0;
2    signal sig2 : integer := 1;
3    sig1 <= 14;
4    sig1 <= sig2;
5    sig1 <= sig2 after 10 ns;
```

3.5.3 Variables

While signals are the external connections between processes, variables are the internal values within a process. They are only used in a sequential manner, unlike the concurrent nature of signals within and between processes. Variables are used within processes and are declared and used as follows:

```
1    variable var1 : integer := 0;
2    variable var2 : integer := 1;
3    var1 := var2;
```

Notice that there is no concept of a delay in the variable assignment; if you need to schedule an event, it is necessary to use a signal.

3.5.4 Boolean Operators

VHDL has a set of standard Boolean operators built in, which are self explanatory. The list of operators are and, or, nand, not, nor, xor. These operators can be applied to BIT, BOOLEAN, or logic types with examples as follows:

```
1    out1 <= in1 and in2;
2    out2 <= in3 or in4;
3    out5 <= not in5;
```

3.5.5 Arithmetic Operators

There are a set of arithmetic operators built into VHDL which again are self explanatory and these are described and examples provided as follows:

Operator	Description	Example
+	Addition	out1 <= in1 + in2;
-	Subtraction	out1 <= in1 - in2;
*	Multiplication	out1 <= in1 * in2;
/	Division	out1 <= in1/in2;
abs	Absolute Value	absin1 <= abs(in1);
mod	Modulus	modin1 <= mod(in1);
rem	Remainder	remin1 <= rem(in1);
**	Exponent	out1 <= in1 ** 3;

3.5.6 Comparison Operators

VHDL has a set of standard comparison operators built in, which are self explanatory. The operators are =, / =, <, <=, >, >=. These operators can be applied to a variety of types as follows:

```
1    in1 < 1
2    in1 /= in2
3    in2 >= 0.4
```

3.5.7 Logical Shifting Functions

VHDL has a set of six built-in logical shift functions which are summarized in the following table:

Operator	Description	Example
sll	Shift Left Logical	reg <= reg sll 2;
srl	Shift Right Logical	reg <= reg srl 2;
sla	Shift Left Arithmetic	reg <= reg sla 2;
sra	Shift Right Arithmetic	reg <= reg sra 2;
rol	Rotate Left	reg <= reg rol 2;
ror	Rotate Right	reg <= reg ror 2;

3.5.8 Concatenation

The concatenation function in VHDL is denoted by the & symbol and is used as follows:

```
1    A <= 1111;
2    B <= 000
3    out1 <= A & B & 1; -- out1 = 11110001;
```

3.6 Decisions and Loops

3.6.1 If-Then-Else

The basic syntax for a simple if statement is as follows:

```
1    if (condition) then
2      -- statements
3    end if;
```

The condition is a Boolean expression, of the form $a > b$ or $a = b$. Note that the comparison operator for equality is a single =, not to be confused with the double == used in some programming languages. For example, if two signals are equal, then setting an output high would be written in VHDL as:

```
1    if ( a = b ) then
2      out1 <= 1;
3    end if;
```

If the decision needs to have both the if and else options, then the statement is extended as follows:

```
1    if (condition) then
2      -- statements
3    else
4      -- statements
5    end if;
```

So in the previous example, we could add the else statements as follows:

```
1    if ( a = b ) then
2      out1 <= 1;
3    else
4      out1 <= 0;
5    end if;
```

And finally, multiple if conditions can be implemented using the general form:

```
1    if (condition1) then
2      -- statements
3    elsif (condition2)
4      -- statements
5      -- more elsif conditions & statements
6    else
7      -- statements
8    end if;
```

With an example:

```
1    if ( a > 10) then
2      out1 <= 1;
3    elsif( a > 5) then
4      out1 <= 0;
```

```
5    else
6      out1 <= 1;
7    end if;
```

3.6.2 Case

As we have seen with the if statement, it is relatively simple to define multiple conditions, but it becomes a little cumbersome, and so the case statement offers a simple approach to branching, without having to use Boolean conditions in every case. This is especially useful for defining state diagrams or for specific transitions between states using enumerated types. An example of a case statement is:

```
1    case testvariable is
2      when 1 =>
3        out1 <= 1;
4      when 2 =>
5        out2 <= 1;
6      when 3 =>
7        out3 <= 1;
8    end case;
```

This can be extended to a range of values, not just a single value:

```
1    case test is
2      when 0 to 4 => out1 <= 1;
```

It is also possible to use Boolean conditions and equations. In the case of the default option (i.e., when none of the conditions have been met), then the term "when others" can be used:

```
1    case test is
2      when 0 => out1 <= 1;
3      when others => out1 <= 0;
4    end case;
```

3.6.3 For

The most basic loop in VHDL is the for loop. This is a loop that executes a fixed number of times. The basic syntax for the for loop is shown below:

```
1    for loopvar in start to finish loop
2      -- loop statements
3    end loop;
```

It is also possible to execute a loop that counts down rather than up, and the general form of this loop is:

```
1    for loopvar in start downto finish loop
2      -- loop statements
3    end loop;
```

A typical example of a for loop would be to pack an array with values bit by bit, for example:

```
1    signal a : std_logic_vector(7 downto 0);
2    for i in 0 to 7 loop
3            a(i) <= 1;
4    end loop;
```

3.6.4 While and Loop

Both the while and loop loops have an indeterminant number of loops, compared to the fixed number of loops in a for loop and as such are usually not able to be synthesized. For FPGA design, they are not feasible as they will usually cause an error when the VHDL model is compiled by the synthesis software.

3.6.5 Exit

The exit command allows a for loop to be exited completely. This can be useful when a condition is reached and the remainder of the loop is no longer required. The syntax for the exit command is shown below:

```
1    for i in 0 to 7 loop
2      if ( i = 4 ) then
3        exit;
4      endif;
5    endloop;
```

3.6.6 Next

The next command allows a for loop iteration to be exited; this is slightly different from the exit command in that the current iteration is exited, but the overall loop continues onto the next iteration. This can be useful when a condition is reached and the remainder of the iteration is no longer required. An example for the next command is shown below:

```
1    for i in 0 to 7 loop
2      if ( i = 4 ) then
3        next;
4      endif;
5    endloop;
```

3.7 Hierarchical Design

3.7.1 Functions

Functions are a simple way of encapsulating behavior in a model that can be reused in multiple architectures. Functions can be defined locally to an architecture or more commonly in a package (discussed in Part 2 of this book), but in this section the basic approach of defining functions will be described. The simple form of a function is to define a header with the input and output variables as shown here:

```
1    function name (input declarations) return output_type is
2      -- variable declarations
3    begin
4      -- function body
5    end
```

For example, a simple function that takes two input numbers and multiplies them together could be defined as follows:

```
1    function mult (a,b : integer) return integer is
2    begin
3         return a * b;
4    end;
```

3.7.2 Packages

Packages are a common single way of disseminating type and function information in the VHDL design community. The basic definition of a package is as follows:

```
1    package name is
2      -- package header contents
3    end package;
4    package body name is
5      -- package body contents
6    end package body;
```

As can be seen, the package consists of two parts, the header and the body. The header is the place where the types and functions are declared, and the package body is where the declarations themselves take place.

For example, a function could be described in the package body and the function is declared in the package header. Take a simple example of a function used to carry out a simple logic function:

```
1    and10 = and(a,b,c,d,e,f,g,h,i,j)
```

The VHDL function would be something like the following:

```
1    function and10 (a,b,c,d,e,f,g,h,i,j : bit) return bit is
2    begin
3      return a and b and c and d and e and f and g and h and i and j;
4    end;
```

The resulting package declaration would then use the function in the body and the function header in the package header thus:

```
1    package new_functions is
2      function and10 (a,b,c,d,e,f,g,h,i,j : bit) return bit;
3    end;
4
5    package body new_functions is
```

```
 6    function and10 (a,b,c,d,e,f,g,h,i,j : bit) return bit is
 7      begin
 8        return a and b and c and d and e and f and g and h and i and j;
 9      end;
10    end;
```

3.7.3 Components

While procedures, functions, and packages are useful in including behavioral constructs generally, with VHDL being used in a hardware design context, often there is a need to encapsulate design blocks as a separate component that can be included in a design, usually higher in the system hierarchy.

The method for doing this in VHDL is called a component. Caution needs to be exercised with components as the method of including components changed radically between VHDL 1987 and VHDL 1993, as such care needs to be taken to ensure that the correct language definitions are used consistently.

Components are a way of incorporating an existing VHDL entity and architecture into a new design without including the previously created model. The first step is to declare the component in a similar way that functions need to be declared. For example, if an entity is called and4, and it has four inputs (a,b,c,d of type bit) and one output (q of type bit), then the component declaration would be of the form shown here:

```
1    component and4
2      port ( a,b,c,d : in bit; q : out bit );
3    end component;
```

Then this component can be instantiated in a netlist form in the VHDL model architecture:

```
1    d1 : and4 port map ( a, b, c, d, q );
```

Note that in this case, there is no explicit mapping between port names and the signals in the current level of VHDL; the pins are mapped in the same order as defined in the component declaration. If each pin is to be defined independently of the order of the pins, then the explicit port map definition needs to be used:

```
1    d1: and4 port map ( a => a, b => b, c=>c, d=>d, q=>q) ;
```

The final thing to note is that this is called the default binding. The binding is the link between the compiled architecture in the current library and the component being used. It is possible, for example, to use different architectures for different instantiated components using the following statement for a single specific device:

```
1    for d1 : and4 use entity work.and4(behavior) port map ( a,b,c,d,q);
```

or the following to specify a specific device for all the instantiated components:

```
1    for all : and4 use entity work.and4(behavior) port map ( a,b,c,d,q);
```

3.7.4 Procedures

Procedures are similar to functions, except that they have more flexibility in the parameters, in that the direction can be in, out or inout. This is useful in comparison to functions where there is generally only a single output (although it may be an array) and avoids the need to create a record structure to manage the return value. Although procedures are useful, they should be used only for small specific functions. Components should be used to partition the design, not procedures, and this is especially true in FPGA design, as the injudicious use of procedures can lead to bloated and inefficient implementations, although the VHDL description can be very compact. A simple procedure to execute a full adder could be of the form:

```
1    procedure full_adder (a,b : in bit; sum, carry : out bit ) is
2    begin
3      sum := a xor b;
4      carry := a and b;
5    end;
```

Notice that the syntax is the same as that for variables (*not* signals), and that multiple outputs are defined without the need for a return statement.

3.8 Debugging Models

3.8.1 Assertions

Assertions are used to check if certain conditions have been met in the model and are extremely useful in debugging models. Some examples:

```
1    assert value <= max_value
2      report Value too large;
3    assert clock_width >= 100 ns
4      report clock width too small;
5      severity failure;
```

3.9 Basic Data Types

3.9.1 Basic Types

VHDL has the following standard types defined as built-in data types:

- bit
- Boolean
- bit_vector
- integer
- real

3.9.2 Data Type: bit

The bit data type is the simple logic type built into VHDL. The type can have two legal values 0 or 1. The elements defined as of type bit can have the standard VHDL built-in logic functions applied to them. Examples of signal and variable declarations of type bit follow:

```
1    signal ina : bit;
2    variable inb : bit := 0;
3    ina <= inb and inc;
4    ind <= 1 after 10 ns;
```

3.9.3 Data Type: Boolean

The Boolean data type is primarily used for decision making, so the test value for if statements is a Boolean type. The elements defined as of type Boolean can have the standard VHDL built-in logic functions applied to them. Examples of signal and variable declarations of type Boolean follow:

```
1    signal test1 : Boolean;
2    variable test2 : Boolean := false;
```

3.9.4 Data Type: Integer

The basic numeric type in VHDL is the integer and is defined as an integer in the range -2147483647 to $+2147483647$. There are obviously implications for synthesis in the definition of integers in any VHDL model, particularly the effective number of bits, and so it is quite common to use a specified range of integer to constrain the values of the signals or variables to within physical bounds. Examples of integer usage follow:

```
1    signal int1 : integer;
2    variable int2: integer := 124;
```

There are two subtypes (new types based on the fundamental type) derived from the integer type which are integer in nature, but simply define a different range of values, as described in the following subsections.

3.9.5 Integer Subtypes: Natural

The Natural subtype is used to define all integers greater than and equal to zero. They are actually defined with respect to the high value of the integer range as follows:

```
1    natural values : 0 to integer'high
```

3.9.6 Integer Subtypes: Positive

The Positive subtype is used to define all integers greater than and equal to one. They are actually defined with respect to the high value of the integer range as follows:

```
1    positive values : 1 to integer'high
```

3.9.7 Data Type: Character

In addition to the numeric types inherent in VHDL, there are also the complete set of ASCII characters available for designers. There is no automatic conversion between characters and a numeric value per se, but there is an implied ordering of the characters defined in the VHDL standard (IEEE Std 1076-1993). The characters can be defined as individual characters or arrays of characters to create strings. The best way to consider characters is as an enumerated type.

3.9.8 Data Type: Real

Floating point numbers are used in VHDL to define real numbers and the predefined floating point type in VHDL is called real. This defines a floating point number in the range $-1.0e38$ to $+10e38$. This is an important issue for many FPGA designs, as most commercial synthesis products do not support real numbers precisely because they are floating point. In practice it is necessary to use integer or fixed point numbers which can be directly and simply synthesized into hardware. An example of defining real signals or variables is shown here:

```
1    signal realno : real;
2    variable realno : real := 123.456;
```

3.9.9 Data Type: Time

Time values are defined using the special time type. These not only include the time value, but also the unit separated by a space. The basic range of the time type value is between -2147483647 to 2147483647 and the basic unit of time is defined as the femtosecond (fs). Each subsequent time unit is derived from this basic unit of the fs as shown here:

```
1    ps = 1000 fs;
2    ns = 1000 ps;
3    us = 1000 ns;
4    ms = 1000 us;
5    min = 60 sec;
6    hr = 60 min;
```

Examples of time definitions are shown here:

```
1    delay : time := 10 ns;
2    wait for 20 us;
3    y <= x after 10 ms;
4    z <= y after delay;
```

3.10 Summary

This chapter provides a very brief introduction to VHDL and is certainly not a comprehensive reference. It enables the reader, hopefully, to have enough knowledge to understand the syntax of the examples in this book. The author strongly recommends that anyone serious about design with VHDL should also obtain a detailed and comprehensive reference book on VHDL, such as Zwolinski [1] (a useful introduction to digital design with VHDL (and a common student textbook)) or Ashenden [3] (a more heavy-duty VHDL reference that is perhaps more comprehensive, but less easy for a beginner to VHDL).

CHAPTER 4

A Verilog Primer: The Essentials

4.1 Introduction

Verilog has been the primary hardware description language (HDL) for digital design
worldwide for probably more than 30 years, but it is only relatively recently that it has begun
to extend beyond its original focus of IC design into the FPGA arena outside the USA.

Verilog as an HDL does have several advantages over other HDLs such as VHDL, as it is both
C-like and also very compact. This makes writing models in Verilog very straightforward for
digital designers with some software background, and fast.

The reduced scale of the syntax (i.e., its compact nature) also makes it less prone to typing
errors simply due to the fewer number of characters often required compared to other
languages.

This chapter will provide a primer for the basics of Verilog, and as for the VHDL primer, the
reader is referred to a large number of textbooks and references (also many online sources) for
more detailed language descriptions and examples. The purpose of this chapter is as a "quick
start" and to provide an overview of the key language features rather than as a full-blown
reference.

4.2 Modules

Verilog is a language that defines the functional blocks using a "module" concept. The basic
principle is that a module will carry out some function (rather like a procedure in C) but the
ultimate goal is that the module will eventually be synthesized into hardware.

A module is defined by the keyword module and endmodule with the general framework as
shown:

```
1    module modulename(list of connections);
2
3    //contents of the model here
4
5    endmodule
```

The top level module that will be simulated will not necessarily have any connections or
parameters.

4.3 Connections

Once we have a basic module definition (with a name) the next step is to connect it up. Consider the example of an 8-bit counter that has a clock (*clk*) and reset (*rst*) input, with the output word defined as *dout* (7 down to 0) The module definition needs to be modified to include the names of the ports in the header as shown here:

```
1    module counter (clk,rst,dout);
2
3    //contents of the model here
4
5    endmodule
```

The module can also be written with the port names spread over several lines such as:

```
1    module counter (
2    clk,
3    rst,
4    dout
5    );
6
7    //contents of the model here
8
9    endmodule
```

The obvious question is "why would we do that?" However, this way makes it simple to add descriptive comments next to each port name definition, which can be extremely helpful in debugging the model.

Comments in Verilog use the // notation (the same as in C++) and so we could write the module header with some additional comments as shown below. As you can see, we can put helpful comments such as the active clock edge (rising or falling), whether the reset signal is active high or low, and finally the width of the dout variable.

```
1    module counter (
2    clk,    // Clock Signal (Rising Edge)
3    rst,    // reset Signal (Active High)
4    dout    // Counter Output (7:0)
5    );
6
7    //contents of the model here
8
9    endmodule
```

The final step at this point of model creation is to define the type of the ports, in terms of direction. For example, are they inputs, outputs or both? Also, what is the width of the bus if they are composite (multiple value) ports? In this simple example, the *clk* and *rst* ports are both inputs, and the *dout* port is an output, of width 8 (tagged with indices 7 down to 0 in classical digital designer format).

This definition is done using the keywords input, output or inout and the expanded port definitions are given here.

```
1    module counter (
2    clk,      // Clock Signal (Rising Edge)
3    rst,      // reset Signal (Active High)
4    dout      // Counter Output (7:0)
5    );
6    // port declarations
7    input clk;
8    input rst;
9    output [7:0] dout;
10
11   //contents of the model here
12
13   endmodule
```

4.4 Wires and Registers

Once a module has been declared and its ports defined, it is then necessary to make those available internally to the module for use as connections. The most basic connection type is called a wire and this is exactly as its name suggests, simply a direct connection.

If we take our counter example, in order to make each port available for internal coding, we need to setup a wire for each input. These are defined using the wire keyword and the extended module is given as follows:

```
1    module counter (
2    clk,      // Clock Signal (Rising Edge)
3    rst,      // reset Signal (Active High)
4    dout      // Counter Output (7:0)
5    );
6    // port declarations
7    input clk;
8    input rst;
9    output [7:0] dout;
10
11   // wire definitions
12   wire clk;
13   wire rst;
14
15   //contents of the model here
16
17   endmodule
```

But what about the counter output? Why did we not simply assign a wire type to the *dout* variable? The answer is that wire is essentially combinatorial (i.e., a direct connection) whereas the output is synchronous and needs to be held in a register. In Verilog we denote this using the reg keyword rather than the simple wire and so the *dout* variable needs to be defined as shown below. Notice that the declaration of the register also needs to have the bus defined in an identical manner to the port declaration.

```
1      module counter (
2      clk,      // Clock Signal (Rising Edge)
3      rst,      // reset Signal (Active High)
4      dout      // Counter Output (7:0)
5      );
6      // port declarations
7      input clk;
8      input rst;
9      output [7:0] dout;
10
11     // wire definitions
12     wire clk;
13     wire rst;
14
15     // Register definitions
16     reg [7:0] dout;
17
18     //contents of the model here
19
20     endmodule
```

4.5 *Defining the Module Behavior*

When we define the module behavior there are several ways in which this can be done. Verilog is in nature a "bottom up" style language and therefore we need to think in terms of assigning signals directly, or acting in terms of hardware "blocks." The two types of behavioral block that we will consider first are the always and initial blocks.

The always block is a way of defining a section of code that will always be activated—in other words the same as a VHDL process, it is a block of hardware that is always active. The initial block, in contrast, is activated on startup, and used to initialize conditions, then never used again.

In order to illustrate how this works in practice we can consider our simple counter. The basic behavior of this module is to count on a rising clock edge, and increment the counter by one, unless the reset is high, in which case the output value will be set to zero.

The basic outline of the always block is as shown here:

```
1      always @ ( posedge clk )   // Count on the Rising Edge of the Clock
2      begin: counter // Start of the Count — block name count
3
4        // Count Code
5
6      end // End of the Count — end of block counter
```

If we look at this code, we can see that the always keyword is used to define the block and that it is activated by the function *posedge*—in other words the rising edge of the *clk* clock signal. The code is contained between a *begin* and *end*, with the name after the begin defining a block name—which is useful in complex designs for the identification of specific signals within blocks.

The final step in the development of our counter is to therefore define the behavior, and implement the check for reset active high and the counter itself.

```
1    module counter (
2    clk,      // Clock Signal (Rising Edge)
3    rst,      // reset Signal (Active High)
4    dout      // Counter Output (7:0)
5    );
6    // port declarations
7    input clk;
8    input rst;
9    output [7:0] dout;
10
11   // wire definitions
12   wire clk;
13   wire rst;
14
15   // Register definitions
16   reg [7:0] dout;
17
18   always @ ( posedge clk )   // Count on the Rising Edge of the Clock
19   begin: COUNTER // Start of the Counter — block name COUNTER
20
21       if (rst == 1'b1) begin
22           dout <= #1 8'b00000000;
23       end
24       else begin
25           dout <= #1 dout + 1;
26       end
27
28   end // End of the Counter — end of block COUNTER
29
30   endmodule
```

4.6 Parameters

The module parameters are defined using the keyword *parameter* with an example given below, where the parameter *buswidth* is defined as a parameter with a default value of 8, and a second parameter *n* is defined as 4.

```
1    module modulename(interface ports);
2    parameter buswidth = 8;
3    parameter n = 4;
4    ...contents of the model here
5
6    endmodule
```

These parameters (*buswidth* and *n*) can then be assigned when the module is instantiated using the following approach:

```
1    moduleinstance #(8,4) modulename(list of ports);
```

If a parameter local to the module is required then we can use the keyword *localparam* to define a parameter that will just be available inside the module.

4.7 Variables

As we have seen with the section on wires and registers, wires are simply connection points, and registers are used to store variables. Therefore, as we have seen previously, registers can be used to store data for the interface section of the model, but also for internal variables.

4.8 Data Types

The basic data type of the register (reg) is a simple digital logic type with four values (0,1,Z,X), with the default value of X. The register data type can then be used as a building block to create more complex types such as buffers of a specific size. For example, to create a register of width 4 bits, with the name *reg*4 use the following syntax:

```
1    reg[3:0] reg4;
```

4.9 Decision Making

The most basic decision-making element is the if statement, and this can be used in a very simple manner as shown in the basic example below, where the variable y is checked and depending on its value, the register *dout* will be assigned the value of a or b.

```
1    if (y == 1)
2      dout = a;
3    else
4      dout = b;
```

If the decision making requires more than a single statement per choice, then use the keywords *begin* and *end* to wrap up several statements such as:

```
1    if (y == 1)
2    begin
3      dout = a;
4      cout = d;
5    end
6    else
7    begin
8      dout = b;
9      cout = e;
10   end
```

Finally, if statements can also be nested to provide multiple options, as can be seen in the following example, where if y=1, then the second nested if checks for the value of the variable x.

```
1      if (y == 1)
2      begin
3        if ( x == 1)
4          dout = a;
5        else
6          cout = d;
7      end
8      else
9      begin
10       dout = b;
11       cout = e;
12     end
```

4.10 Loops

There are two ways of looping in Verilog, *for* and *repeat*. For loops are very similar to C, setting up the start, end and increment values. For example, to set up a loop to carry out a function on each bit of a 32 bit bus, a *for* loop could be used in this way:

```
1      for(i = 0;i<32;i = i + 1)
2        // Complete the function using the loop variable i
3      end
```

The repeat function could be done in a similar way using the *repeat* function, where the loop is handled incrementing the defined number of times, and then it is up to the designer to increment the value (in this case i).

```
1      repeat (32)
2        // Complete the function using the loop variable i
3        i = i + 1;
4      end
```

4.11 Summary

This chapter is a very brief introduction to basic Verilog. For a more comprehensive overview there are many useful references online, such as http://www.asic-world.com/verilog/, or the excellent text by Zwolinski [4]. It should be noted that in this book we are purely considering Verilog, but there is a modern variant called "SystemVerilog" which has more high level and abstract functionality than basic Verilog. However, not all the FPGA tools support SystemVerilog yet, but Verilog is fairly universally usable for any FPGA tool flow.

CHAPTER 5

Design Automation of FPGAs

5.1 Introduction

With the increasing complexity and size of digital designs targeted at FPGAs, the days of hand designing the logic code for hardware at the lowest level (i.e., able to be downloaded directly to the device) have long gone. Designers now must rely on design automation tools to manage the code, syntax checking, simulation, and synthesis functions. In this chapter we will introduce some of the key concepts and highlight how they can be used most effectively.

5.2 Simulation

5.2.1 Simulators

Simulators are a key aspect of the design of FPGAs. They are how we understand the behavior of our designs, check the results are correct, identify syntax errors and even check postsynthesis functionality to ensure that the designs will operate as designed when deployed in a real device. There are a number of simulators available (all the main design automation vendors provide software) and the FPGA vendors also will supply a simulator usually wrapped up with the design flow software for their own devices. This altruistic approach is very useful in learning to use FPGAs as, while they are clearly going to be targeted with libraries at the vendor's devices, the basic simulators tend to be fully featured enough to be very useful for most circumstances. One of the most prevalent simulators is the ModelSim® from Mentor Graphics, which is generally wrapped in with the Altera and Xilinx design software kits and is very commonly used in universities for teaching purposes. It is a general-purpose simulator which can cater for VHDL, Verilog, SystemVerilog and a mixture of all these languages in a general simulation framework. A screen shot of ModelSim in use is shown in Figure 5.1 and it shows the compilation window and a waveform viewer. There are other ways to use the tool including a transcript window and variable lists rather than waveforms—useful in seeing transitions or events.

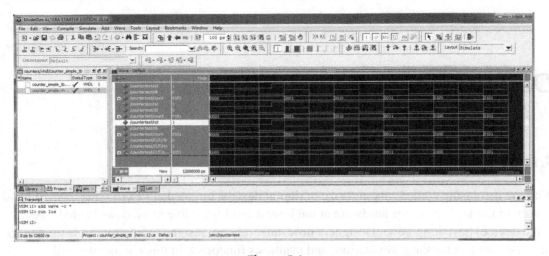

Figure 5.1
ModelSim simulator user interface.

5.2.2 Test Benches

The overall goal of any hardware design is to ensure that the design meets the requirements of the design specification. In order to measure and determine whether this is indeed the case, we need not only to simulate the design representation in a hardware description language (such as VHDL or Verilog), but also to ensure that whatever tests we undertake are appropriate and demonstrate that the specification has been met.

The way that designers can test their designs in a simulator is by creating a *test bench*. This is directly analogous to a real experimental test bench in the sense that stimuli are defined and the responses of the circuit measured to ensure that they meet the specification.

In practice, the test bench is simply a model that generates the required stimuli and checks the responses. This can be in such a way that the designer can view the waveforms and manually check them, or by using appropriate HDL constructs to check the design responses automatically.

5.2.3 Test Bench Goals

The goals of any test bench are twofold. The first is primarily to ensure that correct operation is achieved. This is essentially a functional test. The second goal is to ensure that a synthesized design still meets the specification (particularly with a view to timing errors).

5.2.4 Simple Test Bench: Instantiating Components

Consider a simple model of a VHDL counter given as follows:

```
1   library ieee;
2   use ieee.std_logic_1164.all;
3   use ieee.numeric_std.all;
4
5   entity counter is
6     generic (
7          n : integer := 4
8          );
9     port (
10         clk : in std_logic;
11         rst : in std_logic;
12         output : out std_logic_vector((n-1) downto 0)
13    );
14   end;
15
16   architecture simple of counter is
17   begin
18     process(clk, rst)
19       variable count : unsigned((n-1) downto 0);
20     begin
21       if rst = '0' then
22         count := (others => '0');
23       elsif rising_edge(clk) then
24         count := count + 1;
25       end if;
26       output <= std_logic_vector(count);
27     end process;
28   end;
```

This simple model is a simple counter with a clock and reset signal, and to test the operation of the component we need to do several things.

First, we must include the component in a new design. So we need to create a basic test bench. The listing below shows how a basic entity (with no connections) is created, and then the architecture contains both the component declaration and the signals to test the design.

```
1    library ieee;
2    use ieee.std_logic_1164.all;
3    use ieee.numeric_std.all;
4
5    entity CounterTest is
6    end CounterTest;
7
8    architecture stimulus of CounterTest is
9         signal rst : std_logic := '0';
10        signal clk : std_logic:='0';
11        signal count : std_logic_vector (3 downto 0);
12
13        component counter
14                port(
```

```
15                      clk : in std_logic;
16                      rst : in std_logic;
17                      output : out std_logic_vector(3 downto 0)
18              );
19          end component;
20          for all : counter use entity work.counter ;
21
22      begin
23          DUT: counter port map(clk=>clk,rst=>rst,output=>count);
24          clk <= not clk after 1 us;
25          process
26          begin
27            rst<='0','1' after 2.5 us;
28            wait;
29          end process;
30      end;
```

This test bench will compile in a simulator, and needs to have definitions of the input stimuli (clk and rst) that will exercise the circuit under test (CUT).

If we wish to add stimuli to our test bench we have some significant advantages over our model; the most appealing is that we generally do not need to adhere to any design rules or even make the code synthesizable. Test bench code is generally designed to be off chip and therefore we can make the code as abstract or behavioral as we like and it will still be fit for purpose. We can use wait statements, file read and write, assertions and other nonsynthesizable code options.

5.2.5 Adding Stimuli

In order to add a basic set of stimuli to our test bench, we could simply define the values of the input signals clk and rst with a simple signal assignment:

```
1      clk <= '1';
2      rst <= '0';
```

Clearly this is not a very complex or dynamic test bench, so to add a sequence of events we can modify the signal assignments to include numerous value, time pairs defining a sequence of values.

In our simple example, we wish to go from a reset state to actively counting and so after the rst signal has been initialized to 0, and then set to 1, the clk signal will then increment the counter. In our example we have used a process to set up the clock repetition and a separate process to initialize and then remove the reset.

```
1          clk <= not clk after 1 us;
2          process
3          begin
4            rst<='0','1' after 2.5 us;
5            wait;
6          end process;
```

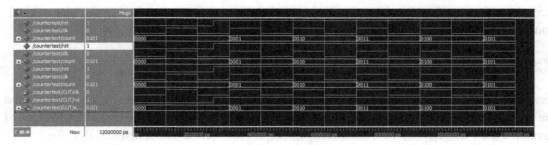

Figure 5.2
ModelSim simulator waveform results.

While this method is useful for small circuits, clearly for more complex realistic designs it is of limited value. Another approach is to define a constant array of values that allow a number of tests to be carried out with a relatively simple test bench and applying a different set of stimuli and responses in turn. The resulting behavior can then be analyzed looking at the waveforms with measurements and markers to ensure the correct timing and functionality, as is shown in Figure 5.2.

For example, we can exhaustively test our simple two-input logic design using a set of data in a record. A record is simply a collection of types grouped together defined as a new type. With a new composite type, such as a record, we can then create an array, just as in any standard type. This requires another type declaration, of the array type itself. With these two new types we can simply declare a constant (of type data_array) that is an array of record values (of type testdata) that fully describe the data set to be used to test the design. Notice that the type data_array does not have a default range, but that this is defined by the declaration in this particular test bench. The beauty of this approach is that we can change from a system that requires every test stimulus to be defined explicitly, to one where a generic test data process will read values from predefined arrays of data. In the simple test example presented here, an example process to apply each set of test data in turn could be implemented as follows:

```
1    process
2    begin
3        for i in test_data'range loop
4            in0 <= test_data(i).in0;
5            in1 <= test_data(i).in1;
6            wait for 100 ns;
7        end loop
8        wait;
9    end process;
```

There are several interesting aspects to this piece of test bench VHDL. The first is that we can use behavioral VHDL (wait for 100 ns) as we are not constrained to synthesize this to hardware. Secondly, by using the range operator, the test bench becomes unconstrained by the size of the data set. Finally, the individual record elements are accessed using the hierarchical construct test_data(i).in0 or test_data(i).in1 respectively.

5.2.6 Assertions

Assertions are a method of mechanically testing values of variables under strict conditions and are prevalent in automated tests of circuits. This is a method by which the timing and values of variables (i.e., signals) can be evaluated against test criteria in a test bench and thus the process of model validation can be automated. This is particularly useful for large designs and also to ensure that any changes do not violate the overall design. It is important to note that assertions are not used in many senses to *debug* a model, but rather to check it is correct in a more formal sense.

For example, we can set an internal variable (flag) to highlight a certain condition (for example, that two outputs have been set to '1'), and then use this with an *assert* statement to report this to the designer and when it has occurred. For example, consider this simple VHDL code:

```
1    assert flag = (x1 and x2)
2      report "The condition has been violated: circuit error"
3      severity Error;
```

If the condition occurs, then during a simulation the assertion will take place and the message reported in the transcript:

```
1    # ** The condition has been violated: circuit error
2    #    Time: 100 ns Iteration: 0 Instance: /testbench
```

This is very useful from a verification perspective as the user can see exactly when the error occurred, but it is a bit cumbersome for debug purposes. The severity of the assertion can be one of note, warning, error, and failure.

The same basic mechanism operates in Verilog, with the same keyword *assert* being used. If we consider the same example, the Verilog code would be something like this:

```
1    assert (x1 & x2) $error( "The condition has been violated: circuit error")
```

In Verilog, the severity command resulting from the assertion can be $fatal, $error (which is the default severity) and $warning.

5.3 Libraries

5.3.1 Introduction

A typical HDL such as Verilog or VHDL as a language on its own is actually very limited in the breadth of the data types and primitive models available. As a result, libraries are required to facilitate design reuse and standard data types for model exchange, reuse, and synthesis. The primary library for standard VHDL design is the IEEE library. Within the IEEE Design Automation Standards Committee (DASC), various committees have developed libraries, packages, and extensions to standard VHDL. Some of these are listed below:

- IEEE Std 1076 Standard VHDL Language
- IEEE Std 1076.1 Standard VHDL Analog and Mixed-Signal Extensions (VHDL-AMS)
- IEEE Std 1076.1.1 Standard VHDL Analog and Mixed-Signal Extensions - Packages for Multiple Energy Domain Support
- IEEE Std 1076.4 Standard VITAL ASIC (Application Specific Integrated Circuit) Modeling Specification (VITAL)
- IEEE Std 1076.6 Standard for VHDL Register Transfer Level (RTL) Synthesis (SIWG)
- IEEE Std 1076.2 IEEE Standard VHDL Mathematical Packages (math)
- IEEE Std 1076.3 Standard VHDL Synthesis Packages (vhdlsynth)
- IEEE Std 1164 Standard Multivalue Logic System for VHDL Model Interoperability (Std_logic_1164)

Each of these working groups consists of volunteers who come from a combination of academia, EDA industry and user communities, and collaborate to produce the IEEE Standards (usually revised every 4 years).

5.3.2 Using Libraries

In order to use a library, first the library must be declared: library ieee; for each library. Within each library a number of VHDL packages are defined, which allow specific data types or functions to be employed in the design. For example, in digital systems design, we require logic data types, and these are not defined in the basic VHDL standard (1076). Standard VHDL defines integer, Boolean, and bit types, but not a standard logic definition. This is obviously required for digital design and an appropriate IEEE standard was developed for this purpose, IEEE 1164. It is important to note that IEEE Std 1164 is NOT a subset of VHDL (IEEE 1076), but is defined for hardware description languages in general.

5.3.3 Std_logic Libraries

In order to use a particular element of a package in a design, the user is required to declare their use of a package using the USE command. For example, to use the standard IEEE logic library, the user needs to add a declaration after the library declaration as follows:

```
1    use ieee.std_logic_1164.all;
```

The std_logic_1164 package is particularly important for most digital designs, especially for FPGA, because it defines the standard logic types used by ALL the commercially available simulation and synthesis software tools and is included as a standard library. It incorporates not only the definition of the standard logic types, but also conversion functions (to and from the standard logic types) and also manages the conversion between signed, unsigned, and logic array variables.

5.4 std_logic Type Definition

As it is such an important type, the std_logic type is described in this section. The type has the following definition:

U	uninitialized. This signal hasn't been set yet.
X	unknown. Impossible to determine this value/result.
0	logic 0
1	logic 1
Z	High Impedance
W	Weak signal, can't tell if it should be 0 or 1
L	Weak signal that should probably go to 0
H	Weak signal that should probably go to 1
-	Don't care

These definitions allow resolution of logic signals in digital designs in a standard manner that is predictable and repeatable across software tools and platforms. The operations that can be carried out on the basic std_logic data types are the standard built-in VHDL logic functions:

- and
- nand
- or
- nor
- xor
- xnor
- not

An example of the use of the std_logic library would be to define a simple logic gate, in this case a three input nand gate as shown in the following listing. The logic pin types are now defined using the IEEE standard library type std_logic rather than the default *bit* type.

```
1    library ieee;
2    use ieee.std_logic_1164.all;
3
4    entity nand3 is
5      port (
6        in0, in1, in2 : in std_logic;
7        out1 : out std_logic
8        );
9    end;
10
```

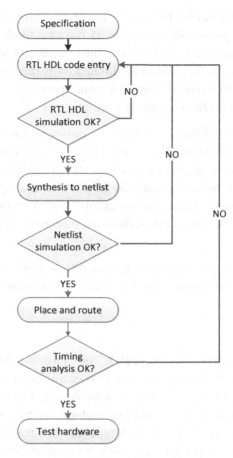

Figure 5.3
HDL based FPGA design flow.

```
11      architecture simple of nand3 is
12      begin
13        out1 <= in0 nand in1 nand in2;
14      end;
```

5.5 Synthesis

5.5.1 Design Flow for Synthesis

The basic HDL design flow is shown in Figure 5.3 and, as can be seen from this figure, synthesis is the key stage between high level design and the physical place and route which is the final product of the design flow. There are several different types of synthesis ranging from behavioral, to RTL and finally physical synthesis.

Behavioral synthesis is the mechanism by which high-level abstract models are synthesized to an intermediate model that is physically realizable. Behavioral models that are not directly synthesizable can be written in VHDL and so care must be taken with high-level models to ensure that this can take place, in fact. There are limited tools that can synthesize behavioral VHDL and these include the Behavioral Compiler from Synopsys, Inc. and MOODS, a research synthesis platform from the University of Southampton.

RTL (Register Transfer Level) synthesis is what most designers call synthesis, and is the mechanism whereby a direct translation of structural and register level VHDL can be synthesized to individual gates targeted at a specific FPGA platform. At this stage, detailed timing analysis can be carried out and an estimate of power consumption obtained. There are numerous commercial synthesis software packages, including Design Compiler®, and Synplify®, but this is not an exhaustive list as there are numerous offerings available at a variety of prices.

Physical synthesis is the last stage in a synthesis design flow and is where the individual gates are placed (using a floor plan) and routed on the specific FPGA platform.

5.5.2 Synthesis Issues

Synthesis basically transforms program-like VHDL into a true hardware design (netlist). It requires a set of inputs, a VHDL description, timing constraints (when outputs need to be ready, when inputs will be ready, data to estimate wire delay), a technology to map to (list of available blocks and their size/timing information) and information about design priorities (area vs. speed). For big designs, the VHDL will typically be broken into modules and then synthesized separately. 10K gates per module was a reasonable size in the 1990s; however, tools can handle a lot more now.

5.6 RTL Design Flow

Register Transfer Level (RTL) VHDL is the input to most standard synthesis software tools. The VHDL must be written in a form that contains registers, state machines (FSM), and combinational logic functions. The synthesis software translates these blocks and functions into gates and library cells from the FPGA library. The RTL design flow is shown in Figure 5.3. Using RTL VHDL restricts the scope of the designer as it precludes algorithmic design, as we shall see later. This approach forces the designer to think at quite a low level, making the resulting code sometimes verbose and cumbersome. It also forces structural decisions early in the design process, which can be restrictive and not always advisable, or helpful.

The design process starts from RTL (Register Transfer Level) VHDL, as follows:

- Simulation (RTL): needed to develop a test bench (VHDL).
- Synthesis (RTL): targeted at a standard FPGA platform.
- Timing Simulation (Structural) simulate to check timing.
- Place and Route using standard tools (e.g., Xilinx Design Manager).

Although there are a variety of software tools available for synthesis (such as LeonardoSpectrum™ or Synplify), they all have generally similar approaches and design flows.

5.7 Physical Design Flow

Synthesis generates a netlist of devices plus interconnections. The Place and Route software figures out where the devices go and how to connect them. The results are not as good as you'd perhaps like: a 40-60% utilization of devices and wires is typical. The designer can trade off run time against greater utilization to some degree, but there are serious limits. Typically the FPGA vendor will provide a software toolkit (such as the Xilinx Design Navigator, or Altera's Quartus® II tools) that manages the steps involved in physical design. Regardless of the particular physical synthesis flow chosen, the steps required to translate the VHDL or EDIF output from an RTL Synthesis software program into a physically downloadable bit file are essentially the same and are listed here:

1. Translate
2. Map
3. Place
4. Route
5. Generate accurate timing models and reports
6. Create binary files for download to device

5.8 Place and Route

There are two main techniques to place and route in current commercial software: which are recursive cut and simulated annealing.

5.8.1 Recursive Cut

In a recursive cut algorithm, we divide the netlist into two halves, and move devices between halves to minimize the number of wires that cross cut (while keeping the number of devices in each half the same). This is repeated to get smaller and smaller blocks.

5.8.2 Simulated Annealing

The simulated annealing method Laarhoven [9] uses a mathematical analogy of the cooling of liquids into solid form to provide an optimal solution for complex problems. The method operates on the principle that annealed solids will find the lowest energy point at thermal equilibrium and this is analogous to the optimal solution in a mathematical problem. The equation for the energy probability used is defined by the Boltzmann distribution given by Equation (5.1):

$$P(E) = \frac{1}{Z(T)} \exp\left(-\frac{E}{k_{\mathrm{B}}T}\right), \tag{5.1}$$

where $Z(T)$ is the partition function, which is a normalization factor dependent on the temperature T, k_{B} is the Boltzmann constant, and E is the energy. This equation is modified into a more general form, as given by (5.2), for use in the simulated annealing algorithm.

$$P(E) = \frac{1}{Q(c)} \exp\left(-\frac{C(i)}{c}\right), \tag{5.2}$$

where $Q(c)$ is a general normalization constant, with a control parameter c, which is analogous to temperature in Equation (5.1). $C(i)$ is the cost function used, which is analogous to the energy in Equation (5.1). The parameters to be optimized are perturbed randomly, within a distribution, and the model tested for improvement. This is repeated with the control parameter decreased to provide a more stable solution. Once the solution approaches equilibrium, then the algorithm can cease. A flowchart of the full algorithm applied is given in Figure 5.4.

5.9 Timing Analysis

Static timing analysis is the most commonly used approach. In static timing analysis, we calculate the delay from each input to each output of all devices. The delays are added up along each path through the circuit to get the critical path through the design and hence the fastest design speed. This works as long as there are no cycles in the circuit; however, in these cases the analysis becomes harder. Design software allows you to break cycles at registers to handle feedback if this is the case. As in any timing analysis, the designer can trade off some accuracy for run time. Digital simulation software such as ModelSim or Verilog will give fast results, but will use approximate models of timing, whereas analog simulation tools like SPICE will give more accurate numbers, but take much longer to run.

5.10 Design Pitfalls

The most common mistake that inexperienced designers make is simply making things too complex. The best approach to successful design is to keep the design elements simple, and the easiest way to manage that is efficient use of hierarchy. The second mistake that is closely

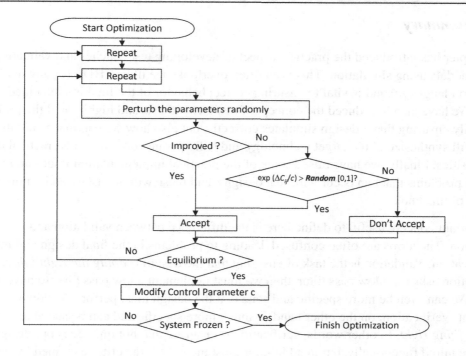

Figure 5.4
Flowchart of the simulated annealing method.

related to design complexity is not testing enough. It is vital to ensure that all aspects of the design are adequately tested. This means not only carrying out basic functional testing, but also systematic testing, and checking for redundant states and potential error states. Another common pitfall is to use multiple clocks unnecessarily. Multiple clocks can create timing-related bugs that are transient or hardware dependent. They can also occur in hardware and yet be missed by simulation.

5.10.1 Initialization

Any default values of signals and variables are ignored. This means that you must ensure that synchronous (or asynchronous) sets and resets must be used on all flip-flops to ensure a stable starting condition. Remember that synthesis tools are basically stupid and follow a basic set of rules that may not always result in the hardware that you expect.

5.10.2 Floating Point Numbers and Operations

Data types using floating point are currently not supported by synthesis software tools. They generally require 32 bits and the requisite hardware is just too large for most FPGA and ASIC platforms.

5.11 Summary

This chapter has introduced the practical aspect of developing test benches and validating VHDL models using simulation. This is an often overlooked skill in VHDL (or any hardware description language) and is vital to ensuring correct behavior of the final implemented design. We have also introduced the concept of design synthesis and highlighted the problem of not only ensuring that a design simulates correctly, but also how we can make sure that the design will synthesize to the target technology and still operate correctly with practical delays and parasitics. Finally, we have raised some of the practical implementation issues and potential problems that can occur with real designs, and these will be discussed in more detail in Part 4 of this book.

An important concept useful to define here is the difference between validation and verification. The terms are often confused, leading to problems in the final design and meeting a specification. Validation is the task of ensuring that the design is *doing the right thing*. If the specification asks for a low pass filter, then we must implement a low pass filter to have a valid design. We can even be more specific and state that the design must perform within a constraint. Verification, on the other hand, is much more specific and can be stated as *doing the right thing right*. In other words, verification is ensuring that not only does our design do what is required functionally, but in addition it must meet ALL the criteria defined by the specification, preferably with some headroom to ensure that the design will operate to the specification under all possible operating conditions.

Finally, we have introduced several design tools and it should be stressed that this in no way indicates a preference, and of course there are a large number of design tools available including some open source tools (many more than I have listed in this chapter). As is usual with the world of design automation, it is really a matter of user preference as to which tools are used.

CHAPTER 6

Synthesis

6.1 Introduction

The original intention of hardware description languages was to have a design specification language for digital circuits. The main goal of the work was to have a design representation that could be simulated to test whether the specification was fit for purpose. When VHDL was standardized as IEEE Standard 1076, the broader application of VHDL for not just simulation but as an integral part of the hardware design flow became possible.

The original method of designing digital circuits was primarily through the development of schematic based designs, using gate libraries to effectively generate RTL netlists directly from the schematics. This is clearly a reasonable technique when the designs are relatively small, but it quickly becomes apparent that for designs of any size this approach is simply not realistic for modern FPGAs that require millions of gates.

EDA companies realized fairly early on in the HDL development process that if there was a standard language that could represent a data flow and a control flow, then the potential existed for automatically generating the gate level HDL from a higher level description, and RTL was the obvious place to start. RTL (Register Transfer Logic) has the advantage of representing the data flow and control flow directly, and can be mapped easily onto standard gate level logic. The resulting synthesis software (such as the Design Compiler from Synopsys) quickly established an important role in the digital design flow for both ASIC and FPGA designs and has in fact proved to be the driving force in the explosion of productivity of digital designers. The modern high density designs would not be possible without RTL synthesis.

For these reasons, modern day designers often simplify RTL synthesis to just "synthesis." However, this is not the whole story. As designs have continued to become more complex, there has been an ever-increasing push to behavioral synthesis; however, there is not the same support from the EDA industry for behavioral synthesis software.

6.1.1 HDL Supported in RTL Synthesis

While VHDL is standardized, synthesis is not, so the VHDL that can be synthesized is a subset of the complete VHDL language. Another common problem for designers is the fact

Design Recipes for FPGAs. http://dx.doi.org/10.1016/B978-0-08-097129-2.00006-4

that different synthesis software packages will give different output results for the same input VHDL, even to the extent that some will synthesize and some will not under certain conditions. This also applies in equal measure to Verilog models, where various constructs will be not be able to be synthesized. Some of these are now discussed in the remainder of this chapter.

There are two types of unsupported elements in VHDL:

- those that will cause a synthesis failure;
- those that are just ignored.

The failure elements are in many respects easier to manage as the synthesis software will provide an error message. It is the ignored elements that can be more insidious as they can obviously leave errors in the synthesized design that may not be picked up until the hardware is tested.

Initial conditions

VHDL supports the initial condition being set for signals and variables; however, this is not physically realized. In practice the initial conditions in the synthesized design are random and so in a practical design a reset condition should always be defined using an external reset pin. This is because, during synthesis, the initial conditions are ignored.

Concurrent edges

It is common to use a clock edge as a trigger for a model, so a simple VHDL model may have a process to wait for the rising edge of a clock.

```
1    process (clk)
2      if rising_edge(clk) then
3        qout <= din;
4      end if;
5    end process;
```

Or in a similar way:

```
1    process (clk)
2      if clk'event and clk='1' then
3        qout <= din;
4      end if;
5    end process;
```

What is NOT valid is to have more than one rising edge as the trigger condition, as this would fail the synthesis.

```
1    process (clk, clk2)
2      if rising_edge(clk) and rising_edge(clk2) then
3        qout <= din;
4      end if;
5    end process;
```

6.2 *Numeric Types*

Synthesis is only supported for numbers that have a finite range. For example, an integer type with an undefined range (infinite) is not supported by synthesis software. In general terms it is often required that designers specify the range of integers and other integer based numbers prior to synthesis (such as signed or unsigned). This can be a subtle restriction as vectors that have a number as the index must have this number defined in advance, so busses cannot be of a variable size. Floating point (real) numbers are generally not supported by synthesis software tools, as they do not have floating point libraries defined.

6.3 *Wait Statements*

Wait statements are only supported if the wait is of the form of an implied sensitivity list and a specific value. So, if the statement is something like:

```
1      Wait on clk = '1';
```

then this is supported for synthesis. If the wait statement is dependent on a specific time delay then this is NOT supported for synthesis. For example, a statement in VHDL such as this is not supported:

```
1      Wait for 10 ns;
```

6.4 *Assertions*

Assertions in any form are ignored by the synthesis software.

6.5 *Loops*

The FOR loop is a special case of the general loop mechanism in VHDL and synthesis requires that the range of the loop must be defined as a static value, globally. This means that you cannot use variables to define the range of the FOR loop *on the fly* for synthesis. If a while loop is implemented, then there has to be a wait statement in the loop somewhere; otherwise, it becomes a potentially infinite loop.

6.6 *Some Interesting Cases Where Synthesis May Fail*

Unfortunately, there are differences between synthesis software packages and so care must be taken to ensure interoperability between packages, particularly in multiteam designs or when using third-party VHDL cores. The cores may have been synthesized using software different from the one you are using in your design flow, so the advertised synthesizable core may not always be synthesizable for you, in your design flow.

Because of this, it is usually a good idea to keep the VHDL as generic as possible and avoid using *tricks* of a particular package if you plan to deliver IP cores or use different tools. This may lead to slightly less compact VHDL, but the reliability of the VHDL will be greater, and potential problems (which could cause significant delays later in the design process, particularly in an integration phase) can be avoided.

One case is the use of different trigger variables in a process. For example, if there is a clock and a reset signal, or a clock and an enable signal, it is tempting to combine the logic into one expression such as:

```
1    if (clk'event and clk='1' and nrst ='0') then
2       ...
3    end if;
```

However, in some synthesis software this would cause an error. It is always preferable to separate these variables into nested if statements for three reasons:

1. The code will be more readable.
2. The chance of undefined logic states is reduced.
3. The synthesis software will not have a problem with your VHDL!

6.7 What Is Being Synthesized?

6.7.1 Overall Design Structure

The basic approach for synthesizing digital circuits is to consider every design block as a combination of a controller and a data path. The controller is generally a finite state machine (FSM), clocked, and the data path is usually combinatorial logic, but there may also be storage in there and so a clock may also be required. The basic outline is shown in Figure 6.1.

6.7.2 Controller

The controller is producing the control signals for the data path logic and may also have external control signals, so there are both internal and external control signals in the general case. As this is a FSM, the design is synchronous and therefore is clocked and will generally have a reset condition.

The controller can be represented using a state diagram or bubble diagram. This shows each individual state and all the transitions between the states. The controller can be of two basic types: Moore (where the output of the state machine is purely dependent on the state variables) and Mealy (where the output can depend on the current state variable values AND the input values). The behavior of the state machine is represented by the state diagram (also sometimes called a state chart) as shown in Figure 6.2.

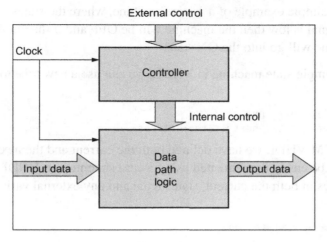

Figure 6.1
Synthesizable digital circuit.

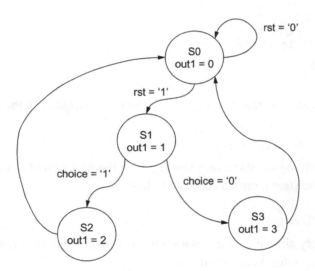

Figure 6.2
Basic finite state machine.

The technique for modeling finite state machines will be covered later in this book, but the key elements to remember are that, as this is a finite state machine, there are a finite number of states, and hence the number of storage elements (D types) is implicit in this definition. Also, the VHDL allows the definition of the state names as an enumerated type, which makes the VHDL readable, easy to understand and also easily synthesizable.

For example, take a simple example of a 2-state machine, where the states are called ON and OFF. If the onoff signal is low then the machine will be OFF and if the onoff switch is high, then the state machine will go into the ON state.

To implement this simple state machine in VHDL, we can use a new type to represent the states:

```
1    type states is (OFF, ON) ;
2    signal current_state, next_state : states;
```

Notice that in the FSM VHDL we have defined both the current and the next state. The main part of the FSM can be easily implemented using a case statement in VHDL within a process that waits for changes in both the current_state signal and any external variables or control signals.

```
1    process (current_state, onoff)
2    begin
3      case current_state is
4      when OFF =>
5        if onoff = '1' then
6          next_state <= ON;
7        end if;
8      when ON =>
9        if onoff = '0' then
10          next_state <= OFF;
11        end if;
12      end case;
13    end process;
```

Elsewhere in the architecture, the current_state needs to be assigned to the next state as follows:

```
1    current_state <= next_state;
```

If we were to do something similar using Verilog, then the first step of defining a new type would be to specify the state names as defined below:

```
1    localparam      ON   = 1'd0;
2    localparam      OFF  = 1'd1;
```

This is obviously a very simple example with only two states, requiring just one bit (4 states would require 2 bits, 8 states 3 and so on).

Using this definition, the state variables can then be defined using a simple register, and again only one bit is required for that:

```
1    reg current_state   = OFF;
2    reg next_state      = OFF;
```

So, the current state and the next state variables have been defined, and the next step is to define the state machine itself. We have several methods to implement this; however, if we stick to the use of a *case* statement, then the following code would demonstrate the correct behavior:

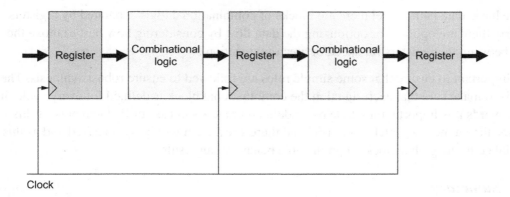

Figure 6.3
Data path.

```
1    always @* begin
2    // Next state logic _____
3      // State machine
4      case (current_state)
5        OFF:    next_state = onoff ? OFF : ON;
6        ON:     next_state = onoff ? ON : OFF;
7        default:        next_state = OFF;
8      endcase
```

As with the VHDL example, the transition to the next state needs to be handled in a synchronous manner, and in Verilog this could be using the rising edge of the clock once again:

```
1    always @ (posedge CLK) begin
2        current_state <= next_state;
3    end
```

6.7.3 Data Path

The data path logic is the logic (as the name suggests) to process the input data and generate the correct output data. The functionality of the data path logic will usually be divided into blocks and this offers the possibility of optimization for speed or area. For example, if area is not an issue, but speed is the primary concern, then a large design could be constructed to generate the output in potentially a single clock cycle. If the area is not an issue, but throughput is required, then pipelining could be used to maximize the overall data rates, although the individual latency may be high. Finally, if area is the critical factor, then single functional blocks can be used and registers used to store intermediate values and the same function applied repeatedly. Clearly this will be a lot slower, but potentially take a lot less space.

In the basic data path model there are blocks of combinational logic separated by registers. Clearly there are options for optimizing the data flow by considering how best to move the data between the registers for speed or area optimization.

It is important to ensure that some simple rules are followed to ensure robust synthesis. The first is to make sure that each signal in the combinational block is defined for every cycle; in other words it is important not to leave undefined branches in case or if statements. If this occurs, then a memory latch is inferred and therefore a latch will be synthesized and as this is not linked to the global clock, unpredictable behavior can result.

6.8 Summary

This chapter has introduced the concept of synthesis, both from a designer's point of view and also the implications of using certain types of VHDL with the intention of synthesizing it. The assumptions and limitations of the various approaches have been described and some sensible practical approaches to obtaining more robust designs defined.

PART 2

Introduction to FPGA Applications

The aim is of the applications part of this book is to identify key points/issues and "nuggets" of information that are of practical use to the designer. The technical information on the issues provided later in the book are referenced, enabling the reader to see the "wood for the trees" and select the "trees" they need to solve a particular issue. Each application uses a combination of block diagrams, state diagrams and code snippets to explain the key concepts in making the application work. Detailed analysis of specific aspects of the design are forward referenced as required.

The first application is a high-speed video monitor system that requires the implementation of a link to a video camera, and also interfaces to RAM and a hard disc. While this is a notional system, the concept is in common usage in a variety of contexts. The techniques covered include the interface to some standard camera formats, handling high speed serial data, managing data storage, and finally integrating the data throughout effectively.

The second application is more about processing power and illustrates the practical aspects of developing multiple processor cores on a standard FPGA platform and how that can be managed in practice. To this end a simple processor is developed from scratch to illustrate how those building blocks can be managed and implemented on an FPGA.

CHAPTER 7

High Speed Video Application

7.1 Introduction

This application is designed to show how several high data rate applications can be handled using VHDL on FPGAs. The system consists of a high-speed camera, processor core, disk drive interface, RAM interface, and serial link to an external PC. The overall system has been chosen to illustrate how to move large amounts of data around quickly and efficiently. The outline of such a test application is shown in the following figure. As can be seen, there are several key aspects involved, but mainly it is about moving large amounts of data around a system quickly, efficiently, and reliably.

The basic system is shown in outline form in Figure 7.1:

The key performance aspect of this system is in the three interfaces:

- Camera to FPGA
- FPGA to PC/Hard Disc Drive (HDD)
- FPGA to RAM

If we consider the basic camera performance criteria, we have four issues to consider:

- Resolution
- Frame rate
- Color specification
- Clip size

In this example, the resolution is defined as being 640 × 480 pixels, the color mode is 24-bit color (3 × 8 bit planes), the maximum frame rate is 100 per second and finally the basic clip size is anything up to 10 s.

What is not shown in the overview figure is the requirement for some basic control options (such as play, record, store) to allow the stored clips to be replayed using a standard VGA output (available on most FPGA development kits) or stored for long-term storage on a hard disc drive (or similar high capacity storage device). This could be handled separately using a PC interface, but that detail is beyond the scope of this basic system description.

Design Recipes for FPGAs. http://dx.doi.org/10.1016/B978-0-08-097129-2.00007-6

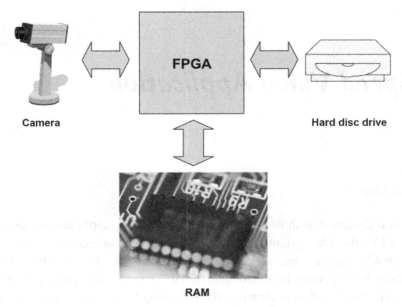

Figure 7.1
Video monitor system overview.

7.2 The Camera Link Interface

7.2.1 Hardware Interface

There are a number of approaches for linking cameras for the high-speed transfer of data, with the two most common being USB (to PCs) and a standard Camera Link using LVDS serial data transmission. The LVDS (Low Voltage Differential Swing) system is a differential serial link that uses voltages of about 350 mV to transmit high-speed data with low noise and low power. Many FPGA development kits have a standard LVDS bus available and this means that the signals can be connected directly between the camera and the FPGA board to transfer data from the camera to the FPGA and hence to the storage (either RAM or HDD).

7.2.2 Data Rates

The actual data rate required is theoretically the resolution multiplied by the frame rate multiplied by the number of bits required for each pixel, which in this example would mean the following calculation:

$$\text{Data rate} = \text{Resolution} * \text{frame rate} * \text{bits/pixel} \qquad (7.1)$$

which for the specification would mean a total data rate of:

$$\text{Data rate} = 640 * 480 * 100 * 24 \qquad (7.2)$$
$$\text{Data rate} = 737,280,000 \, \text{bps} \qquad (7.3)$$

This equates to a data rate of over 90 MB/s (megabytes per second) and as such is extremely fast for a practical application. Even if the FPGA could run at 100 MHz, the margin on such a system is pretty small.

7.2.3 The Bayer Pattern

Luckily, in practice, most camera systems do not use 24 bits in this raw fashion. Kodak has developed the Bayer pattern which is a technique whereby instead of requiring each pixel to have its own individual three color planes (requiring 24 bits in total), an array of color filters is placed over the camera sensor and this limits the requirement for data bits per pixel to a single 8-bit byte (with a known color filter in place). The Bayer pattern is repeated over the image in a fixed configuration to standardize this process. The Bayer pattern is shown in Figure 7.2.

Clearly, using this approach, the required data rate can be divided by three and reduces to a more manageable 30 MB/s. Clearly, the disadvantage of this approach is that the resolution is reduced; however, most images can be reconstructed fairly readily using a method of interpolation which checks firstly which color the current pixel is (red, green, or blue, denoted by R, G or B respectively) and then takes an average of the neighboring pixels of the missing colors. For example, if the current pixel color is green, then the blue and red color of the current pixel is obtained by averaging the neighboring blue (2) and red (2) pixels, respectively.

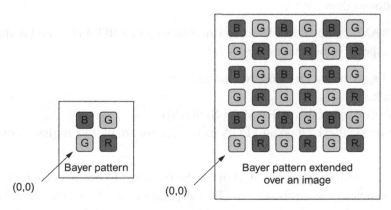

Figure 7.2
Basic Bayer pattern, and extended over a larger image area.

7.2.4 Memory Requirements

Taking the use of Bayer patterns to reduce the sheer amount of data required into account, this means that the RAM requirements are still high; in this case for a 640×480 image size, this will require a memory size of:

$$\text{Memory size} = \text{resolution} * \text{bits/pixel} \tag{7.4}$$

$$\text{Memory size} = \text{resolution} * 8\,\text{bits} \tag{7.5}$$

$$\text{Memory size} = 640 * 480 * 8\,\text{bits} \tag{7.6}$$

$$\text{Memory size} = 307{,}200 * 8\,\text{bits (per frame)} \tag{7.7}$$

Clearly, a large memory is going to be required for any significant memory storage and it is unlikely to be possible to store this on the FPGA itself. A more practical solution will be to use some RAM connected to the FPGA (or perhaps available on the development board itself). Options for the memory could include SDRAM or Flash memory. Both of these options will be discussed in detail later in the book; however, it is useful to consider the advantages and disadvantages of each approach in general. If we consider SDRAM (Synchronous Dynamic Random Access Memory), the key aspects of this type of memory to consider are:

- This type of DRAM (Dynamic RAM) relies on transistor capacitance on gates to store data.
- DRAM is much more compact than SRAM (Static RAM).
- DRAM cannot be synthesized; you need a separate DRAM chip.
- SDRAM requires a synchronization clock that is consistent with the rest of the hardware system (it is designed to operate with microprocessors).
- DRAM data must be refreshed as it is stored charge and decays after a certain time.
- DRAM is slower than SRAM.

Static RAM (SRAM) can be considered in a similar way to a ROM chip and it also has (differing) key aspects of behavior to consider:

- Memory cells are based on standard latches.
- SRAM is fast.
- SRAM is less compact than DRAM (or SDRAM).
- SRAM can be synthesized on an FPGA so is ideal for small, fast registers, or memory blocks.

Static RAM is essentially asynchronous, but can be modified to behave synchronously (as SDRAM is the synchronous equivalent of DRAM), and this is often called Synchronous RAM. Flash memory is useful to consider at this point, even though its operation is fundamentally different from the memory types considered thus far, simply because it is easy to use and is commonly available on many FPGA development boards. Flash memory is essentially a form of EEPROM (electrically programmable ROM) that can be used as a form of persistent RAM. Why persistent? In Flash memory, the device memory is retained even when the power is removed, so it is often used as a form of ROM, which makes it an interesting memory to use on FPGA systems as it could be used to store the FPGA program, but also used as a RAM storage (dynamically) for current data.

7.3 Getting Started

Now that the basic context of the design has been described and the basic specification firmed up, the first stage of the actual design can start. In practice, many of the individual blocks may exist in some form, but may need to be modified to fit the specific application requirements. However, generally speaking it is sensible to start with a top-down design methodology. What that means is that, based on the specification, a top level block can be designed that has the correct pin interface (although this may change as the design is refined) and an outline block structure that contains the functional blocks in the design. If we consider the design example in this part of the book a typical starting point will be a top level diagram showing the basic building blocks of the design and the overall interfaces. Some of the details will not be complete at this stage, but we can start to construct a top level design and we can fill in the details later as we go on with the details of each design block.

Figure 7.3 shows the outline top level design of the application.

The essential features of the design are captured in this sketch: the main functional blocks, the key interfaces and also notice that we have identified a system clock and reset that will propagate to all the individual functional blocks. Notice also that in the original design we did not specify the user input mechanism: that is, how does the user control the camera interface or store data? We have made a design decision at this point, which is to use a simple mouse and keyboard interface to provide the user control to the FPGA system. This allows a flexible approach, so in the first instance, we could use mouse keys or specific keys on the keyboard to initiate a record sequence, or playback, or store, but ultimately, depending on how complex we wish to make the design, it would be possible to design a simple user interface with buttons or similar user interface features, actually on the display to allow controls to drive the system.

7.4 Specifying the Interfaces

From the sketch shown in Figure 7.1 we can begin to identify the interface requirements for the top level design. First, we clearly need a clock and reset (active low), so keeping things simple (always a good strategy) we can define the clock pin as clk and the reset pin as nrst. These are standard logic connections, and so we will use the basic standard logic type defined in the IEEE std_logic library. This does not define any details about the actual implementation

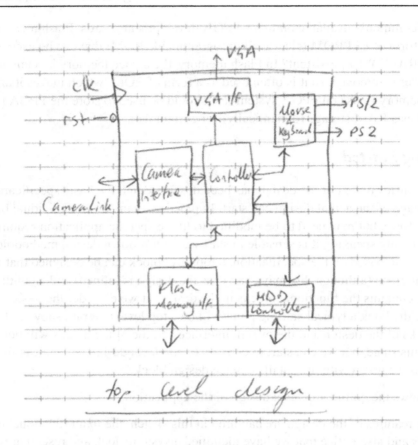

Figure 7.3
Top level design sketch.

of the pins (5V or 3.3V or even 1V), but simply the number of logic levels in the model. The actual implementation is defined by the FPGA being used.

7.5 Defining the Top Level Design

For this design we must define a top level entity name, and also individual block names. It is always a good idea to use meaningful names (unless they become unmanageable, in which case acronyms can be helpful), and hierarchy can also help in keeping duplicate name problems to a minimum. For example, in this case, the design is for an image handler and storage interface, which is clearly a mouthful, so in this example, we will shorten it to IHSI (remember that VHDL is case insensitive). Each main block below this top level will then have the prefix ihsi_ to identify the block in the design. This also has the effect of keeping all the blocks grouped together in the same place alphabetically in the compiled library, which makes things easier to find. We can therefore produce the first initial top level entity for the complete application:

```
1    library ieee;
2    use ieee.std_logic_1164.all;
3    entity ihsi is
4        port (
5            clk : in std_logic;
6            nrst : in std_logic
7        );
8    end entity ihsi;
```

In Verilog this will become:

```
1    module ihsi (clk, nrst)
2    input clk;
3    input nrst;
4
5    endmodule;
```

We can then identify each major block that requires an external interface and add the requisite connection points to the top level entity. It is worth remembering that at each stage of the design, we do not need to have every block defined completely to test other parts of the design. We can use behavioral models or even empty models to simply ensure that the interfaces are in place and then replace each empty block with a fully functional one. We can also start with behavioral models, replace with RTL models and finally even replace these with synthesized ones. Thus, a complete system can be tested piece by piece until all the blocks are in place.

7.6 System Block Definitions and Interfaces

7.6.1 Overall System Decomposition

In this specific application we have several important blocks with external interfaces including:

- Mouse Controller (PS/2)
- Keyboard Controller (PS/2)
- Flash Memory
- VGA Output
- Camera Link
- PC Interface

We can take each of these interfaces in turn and specify the requisite interface connections required for the design.

7.6.2 Mouse and Keyboard Interfaces

The mouse and keyboard PS/2 interfaces are relatively easy. Each of these has a clock and a data connection and so for each we can define two pins as follows:

Mouse: mouse_clk, mouse_data

Keyboard: key_clk, key_data

In the general case, the PS/2 interface (to be covered in more detail in Part 3 of this book) allows both directions to be used (i.e., device to controller and vice versa), so these connections must be defined as INOUT std_logic connections in our top level entity.

7.6.3 Memory Interface

For the memory interface, we have two options. The first option is to define precisely the type of memory we are going to use in this application (RAM, Flash, EEROM, DRAM, SRAM) and produce a specific interface that will work for only that type of memory. Another approach is to consider that we will treat whatever type of memory we have as generic RAM internally, and to design a memory block that will interface to the actual memory—we will treat the memory interface as essentially a virtual RAM block. For the initial design, therefore, we can treat the memory as a simple synchronous RAM block that has a clock, data bus, address bus, and write and read signals. For this initial interface, therefore, we will require the following signals only in VHDL:

Signal	Name	Direction	Type	Notes
Clock	mem_clk	out	std_logic	
Data bus	mem_data(31:0)	inout	std_logic	
Address bus	mem_addr(31:0)	out	std_logic	
Write	mem_nwr	out	std_logic	(active low)
Read	mem_nrd	out	std_logic	(active low)

In Verilog, this will be almost identical, with the definition as follows:

More details on modeling the memory interface and dedicated memory itself is given in Chapter 11.

Signal	Name	Direction	Type	Notes
Clock	mem_clk	out	reg	
Data bus	mem_data	inout	reg [31:0]	
Address bus	mem_addr	out	reg [31:0]	
Write	mem_nwr	out	reg	(active low)
Read	mem_nrd	out	reg	(active low)

7.6.4 The Display Interface: VGA

For the VGA output (to be described later in this book in more detail) we require a specific definition of pins for the connection to the VGA connector on a development board or system. The first set of pins required in any VGA system is the clock and sync pins. The global VGA clock needs to be set to a specific frequency (depending on the monitor), such as 25 MHz, and this must be derived from the system clock on the FPGA board (say 100 MHz). The VGA clock pin is called the pixel clock and we can use the naming convention of vga_ as a prefix, followed by the functional name. So, for the pixel clock, the pin is named vga_out_pixel_clock. In addition to the clock, there are three synchronization signals required, the horizontal sync (vga_hsync), the vertical sync (vga_vsync), and the composite sync (vga_comp_sync). Finally, there is a blank pulse (vga_out_blank_z). The set of pins defined next are the three color data sets. VGA has three color planes (red, green, and blue), each with a definition of 8 bits, giving 24 bits in total. As has been described previously, these can be processed using a Bayer pattern, but when the final output pixel data is put together, all three planes require some output values to be set (even if they are all zero). We can define these pins as 8 bit vectors as follows:

```
1    vga_out_red : out std_logic_vector ( 7 downto 0);
2    vga_out_green : out std_logic_vector (7 downto 0);
3    vga_out_blue : out std_logic_vector (7 downto 0);
```

or in Verilog:

```
1    reg [7:0] vga_out_red;
2    reg [7:0] vga_out_green;
3    reg [7:0] vga_out_blue;
```

This provides a complete definition of the VGA interface to the monitor from the system as a whole. More details of the VGA interface mechanism is given in Chapter 14.

7.7 The Camera Link Interface

The Camera Link standard has been devised to provide a generic 26-pin interface to a wide range of digital cameras and as such we can specify a standard interface at the top level of our design. Although the interface requires 26 pins, they are configured differentially, and so we can specify the basic interface functionally using only 11 pins. There is a clock pin, which we can define as camera_clk, and then four camera control lines defined as cc1 to cc4, respectively. Using the camera_ prefix, we can therefore name these as camera_cc1, camera_cc2, camera_cc3, and camera_cc4. There are two serial communication lines, serTFG (comms to frame grabber) and serTC (comms to camera), which we can name as camera_sertfg and camera_sertc, respectively. Finally, we have the four connection pins from the camera which will contain the data from the device and these are named camera_x0, camera_x1, camera_x2, and camera_x3. Clearly, the actual interface requires differential outputs, and so eventually an extra interface will be required to translate the simple form of interface defined here to the specific pins of the connector.

7.8 The PC Interface

The interface to the PC could be using either a standard serial interface such as USB (covered in Chapter 15) or using a direct interface to a hard disc drive (HDD).

The HDD interface offers a different challenge from the RAM memory interface discussed previously. There are numerous standards for interfacing to HDDs including the major two in current use IDE/AT and SCSI. SCSI (or Small Computers System Interface) is commonly used for high-speed drives and has been historically used extensively in Unix based systems. SCSI is a generic systems interface, and therefore it allows almost ANY type of device to be attached to the system (SCSI) bus.

The IDE/AT standard was devised for HDDs only and so has the advantage of being specifically designed for HDD interfaces. IDE (Intelligent Drive Electronics/AT Attachment) drives are generally slower, but significantly cheaper than SCSI drives and so PCs tend to use an IDE/ATA interface and higher end workstations will use SCSI drives instead.

In this context, the IDE/ATA drive is highly appropriate as the interface is much simpler than the SCSI interface, and therefore more practical in developing a prototype system. If a more advanced system is required, then clearly this can be changed later. The IDE approach is to have a number of master and slave devices on the bus (anyone who has looked inside a PC will recognize the need for setting a master/slave switch or jumper on a drive before installation of an extra or new HDD). A bus controller sets a series of registers with commands and the selected device on the chain will execute. It is worth noting that the bus will operate at the speed of the slowest device on the chain.

There are a total of 13 registers in the IDE/ATA configuration. These registers are divided into command block registers and control block registers. The command block registers are for sending commands to the device or for posting the status of the device. The control block registers are used for device control and for posting an alternate status. The full details of interfacing to an IDE/ATA device is beyond the scope of this book and is not used in this example.

The complexity of the IDE/ATA interface is such that it would probably take several thousand lines of VHDL to implement completely. If the performance requirements were such that it was essential, then the reader can find numerous sources of information to implement this design, including the ATA 6/UDMA100 specification.

An alternative approach is to use a standard interface such as USB with memory buffering and compression to manage the data storage issues, where the USB interface is discussed in detail in Part 3 of this book.

7.9 Summary

In summary, this chapter shows how a high-level specification can be practically decomposed into a series of manageable problems that may all have a relatively simple solution. The key to successful systems design is to decompose the design into blocks that have a definable core function. This can then be implemented directly in VHDL. The second aspect of the design is to analyze the boundaries.

A common phrase coined by systems designers is "problems migrate to the boundaries." In other words, we can easily construct a VHDL design if we know the core functionality; however, getting the individual blocks to communicate successfully is often much harder. As a result, the designer often spends a lot of debug time in integrating a number of different functions together, and being forced to rewrite large sections of code to make that happen.

A useful approach to handling this specific problem is to create empty VHDL models that do not operate functionally, but do have the correct interfaces. These models can be tested with basic communications test data to ensure that the correct signals are in place, the data can be passed around the complete design at the required data rates, and that errors in signal names, directions, and types can be sorted out prior to developing the core VHDL.

This chapter provides a useful introduction to the process of modeling and designing complex systems using VHDL and Verilog. The general approach of thinking at a high level, without going too deeply into the details of each block, has been highlighted.

CHAPTER 8

Simple Embedded Processors

8.1 Introduction

This application example chapter concentrates on the key topic of integrating processors onto FPGA designs. This ranges from simple 8-bit microprocessors up to large IP processor cores that require an element of hardware-software co-design involved. This chapter will take the reader through the basics of implementing a behavioral based microprocessor for evaluation of algorithms, through to the practicalities of structurally correct models that can be synthesized and implemented on an FPGA.

One of the major challenges facing hardware designers in the 21st century is the problem of hardware-software co-design. This has moved on from a basic partitioning mechanism based on standard hardware architectures to the current situation where the algorithm itself can be optimized at a compilation level for performance or power by implementing appropriately at different levels with hardware or software as required. This aspect suits FPGAs perfectly, as they can handle fixed hardware architecture that runs software compiled onto memory, they can implement optimal hardware running at much faster rates than a software equivalent could, and there is now the option of configurable hardware that can adapt to the changing requirements of a modified environment.

8.2 A Simple Embedded Processor

8.2.1 Embedded Processor Architecture

A useful example of an embedded processor is to consider a generic microcontroller in the context of an FPGA platform. Take a simple example of a generic 8-bit microcontroller as shown in Figure 8.1.

As can be seen from Figure 8.1, the microcontroller is a general-purpose microprocessor with a simple clock (clk) and reset (clr), and three 8-bit ports (A, B, and C). Within the microcontroller itself, there needs to be the following basic elements:

Design Recipes for FPGAs. http://dx.doi.org/10.1016/B978-0-08-097129-2.00008-8

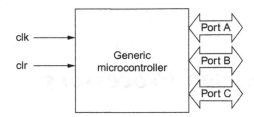

Figure 8.1
Simple microcontroller.

1. A control unit: this is required to manage the clock and reset of the processor, manage the data flow and instruction set flow, and control the port interfaces. There will also need to be a program counter (PC).
2. An ALU: a microcontroller will need to be able to carry out at least some rudimentary processing which is carried out in the ALU (Arithmetic Logic Unit).
3. An Address Bus.
4. A Data Bus.
5. Internal Registers.
6. An instruction decoder.
7. A ROM to hold the program.

While each of these individual elements (1-6) can be implemented simply enough using a standard FPGA, the ROM presents a specific difficulty. If we implement a ROM as a set of registers, then obviously this will be hugely inefficient in an FPGA architecture. However, in most modern FPGA platforms, there are blocks of RAM on the FPGA that can be accessed and it makes a lot of sense to design a RAM block for use as a ROM by initializing it with the ROM values on reset and then using that to run the program.

This aspect of the embedded core raises an important issue, which is the reduction in efficiency of using embedded rather than dedicated cores. There is usually a compromise involved and in this case it is that the ROM needs to be implemented in a different manner, in this case with a hardware penalty. The second issue is what type of memory core to use.

In an FPGA RAM, the memory can usually be organized in a variety of configurations to vary the depth (number of memory addresses required) and the width (width of the data bus). For example, a 512 address RAM block, with an 8-bit address width would be equivalent to a 256 address RAM block with a 16-bit address width.

If the equivalent microcontroller ROM is, say, 12 bits wide and 256, then we can use a 256×16 RAM block and ignore the top 4 bits. The resulting embedded microcontroller core architecture could be of the form shown in Figure 8.2.

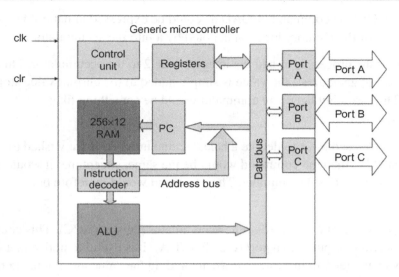

Figure 8.2
Embedded microcontroller architecture.

8.2.2 Basic Instructions

When we program a microprocessor of any type, there are three different ways of representing the code that will run on the processor. These are machine code (1s and 0s), assembler (low level instructions such as LOAD, STORE), and high level code (such as C, Fortran, or Pascal). Regardless of the language used, the code will always be compiled or assembled into machine code at the lowest level for programming into memory. High level code (e.g., C) is compiled and assembler code is assembled (as the name suggests) into machine code for the specific platform.

Clearly a detailed explanation of a compiler is beyond the scope of this book, but the same basic process can be seen in an assembler and this is useful to discuss in this context. Every processor has a basic Instruction Set which is simply the list of functions that can be run in a program on the processor. Take the simple example of the following pseudocode expression:

```
1      b = a + 2;
```

In this example, we are taking the variable a and adding the integer value 2 to it, and then storing the result in the variable b. In a processor, the use of a variable is simply a memory location that stores the value, and so to load a variable we use an assembler command as follows:

```
1    LOAD a
```

What is actually going on here? Whenever we retrieve a variable value from memory, the implication is that we are going to put the value of the variable in the register called the accumulator (ACC). The command "LOAD a" could be expressed in natural language as "LOAD the value of the memory location denoted by a into the accumulator register ACC."

The next stage of the process is to add the integer value 2 to the accumulator. This is a simple matter, as instead of an address, the value is simply added to the current value stored in the accumulator. The assembly language command would be something like:

```
1    ADD #x02
```

Notice that we have used the x to denote a hexadecimal number. If we wished to add a variable, say called c, then the command would be the same, except that it would use the address c instead of the absolute number. The command would therefore be:

```
1    ADD c
```

Now we have the value of a+2 stored in the accumulator register (ACC). This could be stored in a memory location, or put onto a port (e.g., PORT A). It is useful to notice that for a number we use the key character # to indicate that we are adding the value and not using the argument as the address. In the pseudocode example, we are storing the result of the addition in the variable called b, so the command would be something like this:

```
1    STORE b
```

While this is superficially a complete definition of the instruction set requirements, there is one specific design detail that has to be decided on for any processor. This is the number of instructions and the data bus size. If we have a set of instructions with the number of instructions denoted by N, then the number of bits in the opcode (n) must conform to the following rule:

$$N >= 2^n \tag{8.1}$$

In other words, the number of bits provides the number of unique different codes that can be defined, and this defines the size of the instruction set possible. For example, if $n = 3$, then with 3 bits there are 8 possible unique opcodes, and so the maximum size of the instruction set is 8.

8.2.3 Fetch Execute Cycle

The standard method of executing a program in a processor is to store the program in memory and then follow a strict sequence of events to carry out the instructions. The first stage is to use the program counter to increment the program line; this then calls up the next command from memory in the correct order, and then the instruction can be loaded into the appropriate register for execution. This is called the fetch execute cycle.

What is happening at this point? First the contents of the program counter (PC) are loaded into the memory address register (MAR). The data in the memory location are then retrieved and loaded into the memory data register (MDR). The contents of the MDR can then be transferred into the instruction register (IR). In a basic processor, the PC can then be incremented by one (or in fact this could take place immediately after the PC has been loaded into the MDR). Once the opcode (and arguments if appropriate) are loaded, then the instruction can be executed. Essentially, each instruction has its own state machine and control path, which is linked to the instruction register (IR) and a sequencer that defines all the control signals required to move the data correctly around the memory and registers for that instruction. We will discuss registers in the next section, but in addition to the program counter (PC), instruction register (IR) and accumulator (ACC) mentioned already, we require two memory registers at a minimum, the Memory Data Register (MDR) and Memory Address Register (MAR).

For example, consider the simple command LOAD a, from the previous example. What is required to actually execute this instruction? First, the opcode is decoded and this defines that the command is a LOAD command. The next stage is to identify the address. As the command has not used the # symbol to denote an absolute address, this is stored in the variable a. The next stage, therefore, is to load the value in location a into the MDR, by setting MAR = a and then retrieving the value of a from the RAM. This value is then transferred to the accumulator (ACC).

8.2.4 Embedded Processor Register Allocation

The design of the registers partly depends on whether we wish to clone a "real" device or create a modified version that has more custom behavior. In either case there are some mandatory registers that must be defined as part of the design. We can assume that we need an accumulator (ACC), a program counter (PC), and the three input/output ports (PORTA, PORTB, PORTC). Also, we can define the instruction register (IR), Memory Address Register (MAR), Memory Data Register (MDR).

In addition to the data for the ports, we need to have a definition of the port direction and this requires three more registers for managing the tristate buffers into the data bus to and from the ports (DIRA, DIRB, DIRC). In addition to this, we can define a number (essentially arbitrary) of registers for general purpose usage. In the general case the naming, order, and numbering of registers does not matter; however, if we intend to use a specific device as a template, and perhaps use the same bit code, then it is vital that the registers are configured in exactly the same way as the original device and in the same order.

In this example, we do not have a base device to worry about, and so we can define the general purpose registers (24 in all) with the names REG0 to REG23. In conjunction with the general purpose registers, we need to have a small decoder to select the correct register and put the contents onto the data bus (F).

8.2.5 A Basic Instruction Set

In order for the device to operate as a processor, we must define some basic instructions in the form of an instruction set. For this simple example we can define some very basic instructions that will carry out basic program elements, ALU functions, memory functions. These are summarized in the following list of instructions:

*   LOAD arg This command loads an argument into the accumulator. If the argument has the prefix # then it is the absolute number, otherwise it is the address and this is taken from the relevant memory address.
    Examples:
    LOAD #01
    LOAD abc
*   STORE arg This command stores an argument from the accumulator into memory. If the argument has the prefix # then it is the absolute address, otherwise it is the address and this is taken from the relevant memory address.
    Examples:
    STORE #01
    STORE abc
*   ADD arg This command adds an argument to the accumulator. If the argument has the prefix # then it is the absolute number, otherwise it is the address and this is taken from the relevant memory address.
    Examples:
    ADD #01
    ADD abc
*   NOT This command carries out the NOT function on the accumulator.
*   AND arg This command ands an argument with the accumulator. If the argument has the prefix # then it is the absolute number, otherwise it is the address and this is taken from the relevant memory address.
    Examples:
    AND #01
    AND abc
*   OR arg This command ors an argument with the accumulator. If the argument has the prefix # then it is the absolute number, otherwise it is the address and this is taken from the relevant memory address.
    Examples:
    OR #01
    OR abc
*   XOR arg This command xors an argument with the accumulator. If the argument has the prefix # then it is the absolute number, otherwise it is the address and this is taken from the relevant memory address.

Examples:
XOR #01
XOR abc
- INC This command carries out an increment by one on the accumulator.
- SUB arg This command subtracts an argument from the accumulator. If the argument has the prefix # then it is the absolute number, otherwise it is the address and this is taken from the relevant memory address.
  Examples:
  SUB #01
  SUB abc
- BRANCH arg This command allows the program to branch to a specific point in the program. This may be very useful for looping and program flow. If the argument has the prefix # then it is the absolute number, otherwise it is the address and this is taken from the relevant memory address.
  Examples:
  BRANCH #01
  BRANCH abc

In this simple instruction set, there are 10 separate instructions. This implies, from the rule given in equation (8.1) previously in this chapter, that we need at least 4 bits to describe each of the instructions given in the table above. Given that we wish to have 8 bits for each data word, we need to have the ability to store the program memory in a ROM that has words of at least 12 bits wide. In order to cater for a greater number of instructions, and also to handle the situation for specification of different addressing modes (such as the difference between absolute numbers and variables), we can therefore suggest a 16-bit system for the program memory.

Notice that at this stage there are no definitions for port interfaces or registers. We can extend the model to handle this behavior later.

8.2.6 Structural or Behavioral?

So far in the design of this simple microprocessor, we have not specified details beyond a fairly abstract structural description of the processor in terms of registers and busses. At this stage we have a decision about the implementation of the design with regard to the program and architecture.

One option is to take a program (written in assembly language) and simply convert this into a state machine that can easily be implemented in a VHDL model for testing out the algorithm. Using this approach, the program can be very simply modified and recompiled based on simple rules that restrict the code to the use of registers and techniques applicable to the processor in question. This can be useful for investigating and developing algorithms, but is more ideal than the final implementation as there will be control signals and delays due to memory access in a processor plus memory configuration, that will be better in a dedicated hardware design.

Another option is to develop a simple model of the processor that does have some of the features of the final implementation of the processor, but still uses an assembly language description of the model to test. This has advantages in that no compilation to machine code is required, but there are still not the detailed hardware characteristics of the final processor architecture that may cause practical issues on final implementation.

The third option is to develop the model of the processor structurally and then the machine code can be read in directly from the ROM. This is an excellent approach that is very useful for checking both the program and the possible quirks of the hardware/software combination, as the architecture of the model reflects directly the structure of the model to be implemented on the FPGA.

8.2.7 Machine Code Instruction Set

In order to create a suitable instruction set for decoding instructions for our processor, the assembly language instruction set needs to have an equivalent machine code instruction set that can be decoded by the sequencer in the processor. The resulting opcode/instruction table is given here:

Command	Opcode (Binary)
LOAD arg	0000
STORE arg	0001
ADD arg	0010
NOT	0011
AND arg	0100
OR arg	0101
XOR arg	0110
INC	0111
SUB arg	1000
BRANCH arg	1001

8.2.8 Structural Elements of the Microprocessor

Taking the abstract design of the microprocessor given in Figure 8.2 we can redraw with the exact registers and bus configuration as shown in the structural diagram in Figure 8.3. Using

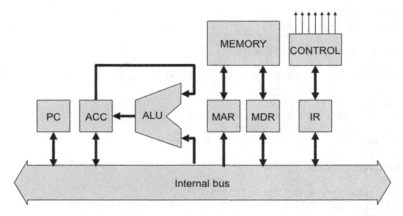

Figure 8.3
Structural model of the microprocessor busses and major blocks.

this model we can create separate VHDL models for each of the blocks that are connected to the internal bus and then design the control block to handle all the relevant sequencing and control flags to each of the blocks in turn. Before this can be started, however, it makes sense to define the basic criteria of the models and the first is to define the basic type. In any digital model (as we have seen elsewhere in this book) it is sensible to ensure that data can be passed between standard models and so in this case we shall use the std_logic_1164 library that is the standard for digital models.

In order to use this library, each signal shall be defined in VHDL of the basic type std_logic and also the library ieee.std_logic_1164.all shall be declared in the header of each of the models in the processor.

Finally, each block in the processor shall be defined as a separate block for implementation in VHDL or Verilog.

8.3 A Simple Embedded Processor Implemented in VHDL

8.3.1 Processor Functions Package

In order to simplify the VHDL for each of the individual blocks, a set of standard functions have been defined in a package called processor_functions. This is used to defined useful types and functions for this set of models. The VHDL for the package is given below:

```
1    library ieee;
2    use ieee.std_logic_1164.all;
3
4    package processor_functions is
5      type opcode is (load, store, add, not, and, or, xor, inc, sub, branch );
6      function decode ( word : std_logic_vector ) return opcode;
7      constant n : integer := 16;
8      constant oplen : integer := 4;
9    type memory_array is array ( 0 to 2**(n-oplen-1) of
10       std_logic_vector(n-1 downto 0);
11      constant reg_zero : unsigned (n-1 downto 0) :=
12      (others => 0);
13   end package processor_functions;
14
15   package body processor_functions is
16     function decode (word : std_logic_vector) return opcode is
17      variable opcode_out : opcode;
18     begin
19      case word(n-1 downto n-oplen-1) is
20        when    0000 => opcode_out := load;
21        when    0001 => opcode_out := store;
22        when    0010 => opcode_out := add;
23        when    0011 => opcode_out := not;
24        when    0100 => opcode_out := and;
25        when    0101 => opcode_out := or;
26        when    0110 => opcode_out := xor;
27        when    0111 => opcode_out := inc;
28        when    1000 => opcode_out := sub;
29        when    1001 => opcode_out := branch;
30        when others => null;
31      end case;
32      return opcode_out;
33     end function decode;
34   end package body processor_functions;
```

8.3.2 The Program Counter

The program counter (PC) needs to have the system clock and reset connections, and the system bus (defined as inout so as to be readable and writable by the PC register block). In addition, there are several control signals required for correct operation. The first is the signal to increment the PC (PC_inc), the second is the control signal to load the PC with a specified value (PC_load) and the final is the signal to make the register contents visible on the internal bus (PC_valid). This signal ensures that the value of the PC register will appear to be high impedance (Z) when the register is not required on the processor bus. The system bus (PC_bus) is defined as a std_logic_vector, with direction inout to ensure the ability to read and write. The resulting VHDL entity is given here:

```
1    library ieee;
2    use ieee.std_logic_1164.all;
3    entity pc is
4      port (
```

```
5        clk : in std_logic;
6        nrst : in std_logic;
7        pc_inc : in std_logic;
8        pc_load : in std_logic;
9        pc_valid : in std_logic;
10       pc_bus : inout std_logic_vector(n-1 downto 0)
11       );
12    end entity pc;
```

The architecture for the program counter must handle all of the various configurations of the program counter control signals and also the communication of the data into and from the internal bus correctly. The PC model has an asynchronous part and a synchronous section. If the PC_valid goes low at any time, the value of the PC_bus signal should be set to Z across all of its bits. Also, if the reset signal goes low, then the PC should reset to zero.

The synchronous part of the model is the increment and load functionality. When the clk rising edge occurs, then the two signals PC_load and PC_inc are used to define the function of the counter. The precedence is that if the increment function is high, then regardless of the load function, the counter will increment. If the increment function (PC_inc) is low, then the PC will load the current value on the bus, if and only if the PC_load signal is also high. The resulting VHDL is given as:

```
1     architecture rtl of pc is
2       signal counter : unsigned (n-1 downto 0);
3     begin
4       pc_bus <= std_logic_vector(counter)
5     when pc_valid = 1 else (others => z);
6       process (clk, nrst) is
7       begin
8         if nrst = 0 then
9           count <= 0;
10        elsif rising_edge(clk) then
11          if pc_inc = 1 then
12            count <= count + 1;
13          else
14            if pc_load = 1 then
15              count <= unsigned(pc_bus);
16            end if;
17          end if;
18        end if;
19      end process;
20    end architecture rtl;
```

8.3.3 The Instruction Register

The instruction register (IR) has the same clock and reset signals as the PC, and also the same interface to the bus (IR_bus) defined as a std_logic_vector of type INOUT. The IR also has two further control signals, the first being the command to load the instruction register (IR_load), and the second being to load the required address onto the system bus (IR_address). The final connection is the decoded opcode that is to be sent to the system controller. This is defined as a simple unsigned integer value with the same size as the basic system bus. The basic VHDL for the entity of the IR is given as follows:

```
1    library ieee;
2    use ieee.std_logic_1164.all;
3    use work.processor_functions.all;
4    entity ir is
5      port (
6        clk : in std_logic;
7        nrst : in std_logic;
8        ir_load : in std_logic;
9        ir_valid : in std_logic;
10       ir_address : in std_logic;
11       ir_opcode : out opcode;
12       ir_bus : inout std_logic_vector(n-1 downto 0)
13     );
14   end entity ir;
```

The function of the IR is to decode the opcode in binary form and then pass to the control block. If the IR_valid is low, the the bus value should be set to Z for all bits. If the reset signal (nsrt) is low, then the register value internally should be set to all 0s.

On the rising edge of the clock, the value on the bus shall be sent to the internal register and the output opcode shall be decoded asynchronously when the value in the IR changes. The resulting VHDL architecture is given here:

```
1    architecture rtl of ir is
2
3      signal ir_internal : std_logic_vector (n-1 downto 0);
4    begin
5      ir_bus <= ir_internal
6    when ir_valid = 1 else (others => z);
7      ir_opcode <= decode(ir_internal);
8      process (clk, nrst) is
9      begin
10       if nrst = 0 then
11         ir_internal <= (others => 0);
12       elsif rising_edge(clk) then
13         if ir_load = 1 then
14           ir_internal <= ir_bus;
15         end if;
16       end if;
17     end process;
18   end architecture rtl;
```

In this VHDL, notice that we have used the predefined function Decode from the processor_functions package previously defined. This will look at the top 4 bits of the address given to the IR and decode the relevant opcode for passing to the controller.

8.3.4 The Arithmetic and Logic Unit

The Arithmetic and Logic Unit (ALU) has the same clock and reset signals as the PC, and also the same interface to the bus (ALU_bus) defined as a std_logic_vector of type INOUT. The ALU also has three further control signals, which can be decoded to map to the eight individual functions required of the ALU. The ALU also contains the Accumulator (ACC) which is a std_logic_vector of the size defined for the system bus width. There is also a single bit output ALU_zero which goes high when all the bits in the accumulator are zero. The basic VHDL for the entity of the ALU is given as follows:

```
1    library ieee;
2    use ieee.std_logic_1164.all;
3    use work.processor_functions.all;
4    entity alu is
5      port (
6        clk : in std_logic;
7        nrst : in std_logic;
8        alu_cmd : in std_logic_vector(2 downto 0) ;
9        alu_zero : out std_logic;
10       alu_valid : in std_logic;
11       alu_bus : inout std_logic_vector(n−1 downto 0)
12     );
13   end entity alu;
```

The function of the ALU is to decode the ALU_cmd in binary form and then carry out the relevant function on the data on the bus, and the current data in the accumulator. If the ALU_valid is low, then the bus value should be set to Z for all bits. If the reset signal (nsrt) is low, then the register value internally should be set to all 0s. On the rising edge of the clock, the value on the bus shall be sent to the internal register and the command shall be decoded. The resulting VHDL architecture is given here:

```
1    architecture rtl of alu is
2      signal acc : std_logic_vector (n−1 downto 0);
3    begin
4      alu_bus <= acc
5    when acc_valid = 1 else (others => z);
6      alu_zero <= 1 when acc = reg_zero else 0;
7      process (clk, nrst) is
8      begin
9        if nrst = 0 then
10         acc <= (others => 0);
11       elsif rising_edge(clk) then
12         case acc_cmd is
13         -- load the bus value into the accumulator
14         when 000 => acc <= alu_bus;
15         -- add the acc to the bus value
16         when 001 => acc <= add(acc,alu_bus);
17         -- not the bus value
18         when 010 => acc <= not alu_bus;
19         -- or the acc to the bus value
20         when 011 => acc <= acc or alu_bus;
```

```
21          -- and the acc to the bus value
22          when 100 => acc <= acc and alu_bus;
23          -- xor the acc to the bus value
24          when 101 => acc <= acc xor alu_bus;
25          -- increment acc
26          when 110 => acc <= acc + 1;
27          -- store the acc value
28          when 111 => alu_bus <= acc;
29        end if;
30      end process;
31    end architecture rtl;
```

8.3.5 The Memory

The processor requires a RAM memory, with an address register (MAR) and a data register (MDR). There therefore needs to be a load signal for each of these registers: MDR_load and MAR_load. As it is a memory, there also needs to be an enable signal (M_en), and also a signal to denote Read or Write modes (M_rw). Finally, the connection to the system bus is a standard inout vector as has been defined for the other registers in the microprocessor.

The basic VHDL for the entity of the memory block is given here:

```
1    library ieee;
2    use ieee.std_logic_1164.all;
3    use work.processor_functions.all;
4    entity memory is
5      port (
6        clk : in std_logic;
7        nrst : in std_logic;
8        mdr_load : in std_logic;
9        mar_load : in std_logic;
10       mar_valid : in std_logic;
11       m_en : in std_logic;
12       m_rw : in std_logic;
13       mem_bus : inout std_logic_vector(n-1 downto 0)
14       );
15     end entity memory;
```

The memory block has three aspects. The first is the function in which the memory address is loaded into the memory address register (MAR). The second function is either reading from or writing to the memory using the memory data register (MDR). The final function, or aspect, of the memory is to store the actual program that the processor will run. In the VHDL model, we will achieve this by using a constant array to store the program values.

The resulting basic VHDL architecture is given as follows:

```
1
2    architecture rtl of memory is
3      signal mdr : std_logic_vector(wordlen-1 downto 0);
4      signal mar : unsigned(wordlen-oplen-1 downto 0);
5    begin
6      mem_bus <= mdr
```

```
 7        when mem_valid = 1 else (others => z);
 8          process (clk, nrst) is
 9            variable contents : memory_array;
10            constant program : contents :=
11            (
12              0 => 0000000000000011,
13              1 => 0010000000000100,
14              2 => 0001000000000101,
15              3 => 0000000000001100,
16              4 => 0000000000000011,
17              5 => 0000000000000000 ,
18              others => (others => 0)
19            );
20          begin
21            if nrst = 0 then
22              mdr <= (others => 0);
23              mdr <= (others => 0);
24              contents := program;
25            elsif rising_edge(clk) then
26              if mar_load = 1 then
27                mar <= unsigned(mem_bus(n-oplen-1 downto 0));
28              elsif mdr_load = 1 then
29                mdr <= mem_bus;
30              elsif mem_en = 1 then
31                if mem_rw = 0 then
32                  mdr <= contents(to_integer(mar));
33                else
34                  mem(to_integer(mar)) := mdr;
35                end if;
36              end if;
37            end if;
38          end process;
39        end architecture rtl;
```

We can look at some of the VHDL in a bit more detail and explain what is going on at this stage. There are two internal signals to the block, mdr and mar (the data and address, respectively). The first aspect to notice is that we have defined the MAR as an unsigned rather than as a std_logic_vector. We have done this to make indexing direct. The MDR remains as a std_logic_vector. We can use an integer directly, but an unsigned translates easily into a std_logic_vector.

```
1        signal mdr : std_logic_vector(wordlen-1 downto 0);
2        signal mar : unsigned(wordlen-oplen-1 downto 0);
```

The second aspect is to look at the actual program itself. We clearly have the possibility of a large array of addresses, but in this case we are defining a simple three line program:

```
1        c = a + b
```

The binary code is shown below:

```
1            0 => 0000000000000011
2            1 => 0010000000000100
3            2 => 0001000000000101
4            3 => 0000000000001100
```

```
5          4 => 0000000000000011
6          5 => 0000000000000000
7          Others => (others => 0)
```

For example, consider the line of the declared value for address 0. The 16 bits are defined as 0000000000000011. If we split this into the opcode and data parts we get the following:

```
1    Opcode 0000
2    Data   000000000011
```

In other words, this means LOAD the variable from address 3. Similarly, the second line is ADD from 4, finally the third command is STORE in 5. In addresses 3, 4, and 5, the three data variables are stored.

8.3.6 Microcontroller Controller

The operation of the processor is controlled in detail by the sequencer, or controller block. The function of this part of the processor is to take the current program counter address, look up the relevant instruction from memory, move the data around as required, setting up all the relevant control signals at the right time, with the right values. As a result, the controller must have the clock and reset signals (as for the other blocks in the design), a connection to the global bus, and finally all the relevant control signals must be output. An example entity of a controller is given here:

```
1    library ieee;
2    use ieee.std_logic_1164.all;
3    use work.processor_functions.all;
4    entity controller is
5      generic (
6        n : integer := 16
7      );
8      port (
9        clk : in std_logic;
10       nrst : in std_logic;
11       ir_load : out std_logic;
12       ir_valid : out std_logic;
13       ir_address : out std_logic;
14       pc_inc : out std_logic;
15       pc_load : out std_logic;
16       pc_valid : out std_logic;
17       mdr_load : out std_logic;
18       mar_load : out std_logic;
19       mar_valid : out std_logic;
20       m_en : out std_logic;
21       m_rw : out std_logic;
22       alu_cmd : out std_logic_vector(2 downto 0);
23       control_bus : inout std_logic_vector(n-1 downto 0)
24     );
25   end entity controller;
```

Using this entity, the control signals for each separate block are then defined, and these can be used to carry out the functionality requested by the program. The architecture for the

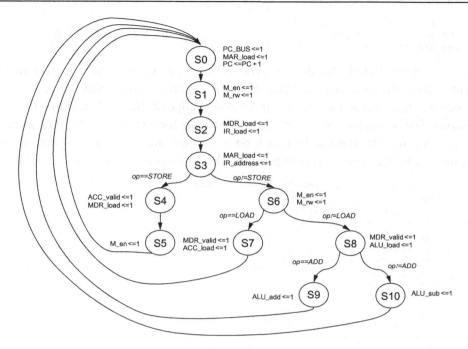

Figure 8.4
Basic processor controller state machine.

controller is then defined as a basic state machine to drive the correct signals. The basic state machine for the processor is defined in Figure 8.4.

We can implement this using a basic VHDL architecture that implements each state using a new state type and a case statement to manage the flow of the state machine. The basic VHDL architecture follows and it includes the basic synchronous machine control section (reset and clock) and the management of the next stage logic.

```
1    architecture rtl of controller is
2      type states is (s0,s1,s2,s3,s4,s5,s6,s7,s8,s9,s10);
3      signal current_state, next_state : states;
4    begin
5        state_sequence: process (clk, nrst) is
6          if nrst = 0 then
7            current_state <= s0;
8          else
9            if rising_edge(clk) then
10             current_state <= next_state;
11           end if;
12         end if;
13       end process state_sequence;
14
15       state_machine : process ( present_state, opcode ) is
```

```
16        -- state machine goes here
17        end process state_machine;
18        end architecture;
```

You can see from this VHDL that the first process (state_sequence) manages the transition of the current_state to the next_state and also the reset condition. Notice that this is a synchronous machine and as such waits for the rising_edge of the clock, and that the reset is asynchronous. The second process (state_machine) waits for a change in the state or the opcode and this is used to manage the transition to the next state, although the actual transition itself is managed by the state_sequence process. This process is given in the VHDL here:

```
1         state_machine : process ( present_state, opcode ) is
2         begin
3           -- reset all the control signals
4           ir_load <= 0 ;
5           ir_valid <= 0 ;
6           ir_address <= 0 ;
7           pc_inc <= 0 ;
8           pc_load <= 0 ;
9           pc_valid <= 0 ;
10          mdr_load <= 0 ;
11          mar_load <= 0 ;
12          mar_valid <= 0 ;
13          m_en <= 0 ;
14          m_rw <= 0 ;
15          case current_state is
16            when s0 =>
17            pc_valid<= 1 ;
18            mar_load<= 1 ;
19            pc_inc<= 1 ;
20            pc_load<= 1 ;
21            next_state<=s1;
22            when s1 =>
23            m_en<= 1 ;
24            m_rw<= 1 ;
25            next_state<=s2;
26            when s2 =>
27            mdr_valid<= 1 ;
28            ir_load<= 1 ;
29            next_state<=s3;
30            when s3 =>
31            mar_load<= 1 ;
32            ir_address<= 1 ;
33            if opcode = store then
34              next_state<=s4;
35            else
36              next_state <=s6;
37            end if;
38            when s4 =>
39            mdr_load<= 1 ;
40            acc_valid<= 1 ;
41            next_state<=s5;
42            when s5 =>
43            m_en <= 1 ;
```

```
44            next_state <= s0;
45          when s6 =>
46          m_en<= 1 ; m_rw<= 1 ;
47          if opcode = load then
48            next_state<=s7;
49          else
50            next_state <=s8;
51          end if;
52          when s7 =>
53          mdr_valid<= 1 ;
54          acc_load<= 1 ;
55          next_state<=s0;
56          when s8 =>
57          m_en<= 1 ;
58          m_rw<= 1 ;
59          if opcode = add then
60            next_state<=s9;
61          else
62            next_state <=s10;
63          end if;
64          when s9 =>
65          alu_add <= 1 ;
66          next_state<=s0;
67          when s10 =>
68          alu_sub <= 1 ;
69          next_state<=s0;
70        end case;
71    end process state_machine;
```

8.3.7 Summary of a Simple Microprocessor Implemented in VHDL

Now that the important elements of the processor have been defined, it is a simple matter to instantiate them in a basic VHDL netlist and create a microprocessor using these building blocks. It is also a simple matter to modify the functionality of the processor by changing the address/data bus widths or extend the instruction set.

8.4 A Simple Embedded Processor Implemented in Verilog

As in the case of the VHDL model we can implement common functions in a series of Verilog files for use in the key blocks of the processor. The architecture has been implemented in a slightly different manner, to illustrate a different approach. In both cases an internal bus has been used, which is analogous to the approach taken in early processors; however, a more direct approach can also be taken where the internal registers are accessed directly.

8.4.1 The Program Counter

The program counter (PC) needs to have the system clock and reset connections, and the system bus (defined as inout so as to be readable and writable by the PC register block). In addition, there are several control signals required for correct operation. The first is the signal to increment the PC (pc_inc), the second is the control signal to load the PC with a specified value (pc_load) and the final is the signal to make the register contents visible on the internal bus (pc_valid). This signal ensures that the value of the PC register will appear to be high impedance (Z) when the register is not required on the processor bus. The pc value output (pc_bus) is defined as a standard logic type, with direction inout to ensure the ability to read from and write to the bus.

The architecture for the program counter must handle all of the various configurations of the program counter control signals and also the communication of the data into and from the internal bus correctly. The PC model has an asynchronous part and a synchronous section. If the pc_valid goes low at any time, the value of the pc_bus signal should be set to Z across all of its bits. Also, if the reset signal goes low, then the PC should reset to zero.

The synchronous part of the model is the increment and load functionality. When the clk rising edge occurs, then the two signals pc_load and pc_inc are used to define the function of the counter. The precedence is that if the increment function is high, then regardless of the load function, the counter will increment. If the increment function (pc_inc) is low, then the PC will load the current value on the bus, if and only if the pc_load signal is also high.

The resulting Verilog code is given below:

```
1    'define N 8
2
3    module pc (clk,nrst,pc_inc,pc_valid,pc_load,data);
4
5    input clk;
6    input nrst;
7    input pc_inc;
8    input pc_valid;
9    input pc_load;
10
11   inout ['N-1:0] data;
12
13   wire ['N-1:0] data;
14
15   reg ['N-1:0] counter;
16
17   assign data = pc_valid ? counter : 'N'bz;
18
19   always @(posedge clk) begin
20     if(nrst==0) begin
21       counter <= 0;
22     end
23     else begin
24       if(pc_inc==1) begin
25         counter <= counter + 1;
26       end
27       else begin
```

```
28        if(pc_load==1) begin
29          counter <= data;
30        end
31        else begin
32          counter <= 0;
33        end
34      end
35    end
36  end
37
38  endmodule
```

We can test this using a test bench that first resets the program counter (PC) to initialize it, increments, then loads in a set value (in this case 4), resets the counter and finally sets the valid signal to low so as to disable the output. This is shown in the following test bench Verilog:

```
1   'define N 8
2
3
4   module pc_tb();
5   // declare the counter signals
6   reg clk;
7   reg nrst;
8   reg pc_valid;
9   reg pc_load;
10  reg pc_inc;
11  wire ['N-1:0] data;
12  reg ['N-1:0] datareg;
13
14  // Set up the initial variables and reset
15  initial begin
16    $display ("time\t clk reset inc load valid data");
17    $monitor ("%g\t %b %b %b %b %b %b",
18    $time, clk, nrst, pc_inc, pc_load, pc_valid, data);
19    clk = 1;      // initialize the clock to 1
20    nrst = 1;     // set the reset to 1 (not reset)
21    pc_valid = 0;
22    pc_inc =0;
23    pc_load = 0;
24    datareg = 4;
25    #5 nrst = 0;  // reset = 0 : resets the counter
26    #10 nrst = 1; // reset back to 1 : counter can start
27    #10 pc_inc = 1;
28    #10 pc_inc = 0;
29    #10 pc_load = 1;
30    #10 datareg = 8'bzzzzzzzz;
31    #10 pc_load = 0;
32    #10 pc_inc = 1;
33    pc_valid = 1;
34    #50 pc_valid = 0; // reset back to 1 : counter can start
35    #200 $finish; // Finish the simulation
36  end
37
38  // Clock generator
39  always begin
```

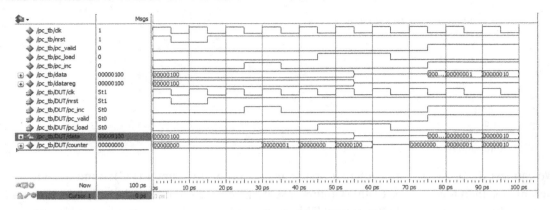

Figure 8.5
Basic processor PC simulation.

```
40       #5 clk = ~clk; // Clock every 5 time slots
41    end
42
43    assign data = datareg;
44
45    // Connect DUT to test bench
46    pc DUT ( clk,nrst,pc_inc,pc_valid,pc_load,data);
47
48    endmodule
```

The resulting waveform shows the behavior as predicted (Figure 8.5):

8.4.2 The Instruction Register

The Instruction Register (IR) in a simple microprocessor is a simple register with enough bits for the address and opcode combined. For example, if the address requires 8 bits, and the opcode also requires 8 bits, then the Instruction Register needs to be 16 bits wide (8 + 8). If the output from the Memory Data Register goes onto the main bus, then this can be read into the instruction register, which is 16 bits wide in this case.

The current value of the instruction register also can be read by the Memory Address Register (MAR) or the Program Counter (PC) and so the stored value needs to be of type inout, so that it can be made valid onto the internal system bus.

The instruction register (IR) therefore has clock and reset signals, and also the same interface to the internal processor bus (ir_bus) defined as a standard logic of direction inout. The IR also has two further control signals, the first being the command to load the instruction register (ir_load), and the second being to make the required address available on the system bus (ir_valid). This consists of the opcode and address, which can be used by the controller or Program Counter.

The code for the Instruction Register is therefore given in the following listing:

```
1    'define OP 8
2    'define ADDR 8
3
4    module ir (clk,nrst,ir_valid,ir_load,ir_bus);
5
6    input clk;
7    input nrst;
8    input ir_valid;
9    input ir_load;
10
11   inout ['OP+'ADDR-1:0] ir_bus;
12
13   wire ['OP+'ADDR-1:0] ir_bus;
14
15   reg ['OP+'ADDR-1:0] ir_reg;
16
17   assign ir_bus = ir_valid ? ir_reg : 16'bz;
18
19   always @(posedge clk) begin
20     if(nrst==0) begin
21       ir_reg <= 0;
22     end
23     else begin
24       if(ir_load==1) begin
25         ir_reg <= ir_bus;
26       end
27     end
28   end
29
30   endmodule
```

We can test this by loading in a sample instruction, and then setting it valid so that it is then seen on the bus. The test bench to achieve this is shown here:

```
1    'define OP 8
2    'define ADDR 8
3
4
5    module ir_tb();
6    // declare the counter signals
7    reg clk;
8    reg nrst;
9    reg ir_valid;
10   reg ir_load;
11
12   wire ['OP+'ADDR-1:0] data;
13   reg ['OP+'ADDR-1:0] datareg;
14
15   // Set up the initial variables and reset
16   initial begin
17     $display ("time\t clk reset inc load valid data");
18     $monitor ("%g\t %b %b %b %b %b",
19     $time, clk, nrst, ir_load, ir_valid, data);
20     clk = 1;    // initialize the clock to 1
```

```
21      nrst = 1;    // set the reset to 1 (not reset)
22      ir_valid = 0;
23      ir_load = 0;
24      datareg = 16'b0000000000001111;
25      #5 nrst = 0; // reset = 0 : resets the counter
26      #10 nrst = 1; // reset back to 1 : counter can start
27      #10 ir_load = 1;
28      #10 ir_load = 0;
29      #10 datareg = 16'bzzzzzzzzzzzzzzzz;
30      ir_valid = 1;
31      #50 ir_valid = 0; // reset back to 1 : counter can start
32      #200 $finish; // Finish the simulation
33    end
34
35    // Clock generator
36    always begin
37      #5 clk = ~clk; // Clock every 5 time slots
38    end
39
40    assign data = datareg;
41
42    // Connect DUT to test bench
43    ir DUT ( clk,nrst,ir_valid,ir_load,data);
44
45    endmodule
```

The resulting waveform shows the behavior as predicted (Figure 8.6):

8.4.3 Memory Data Register

The memory data register is used to handle the data transferred to and from the memory unit, and this can be handled either using a bus approach (which we have used in this architecture) or separate data input and output declaration for the memory. In this case we will use a

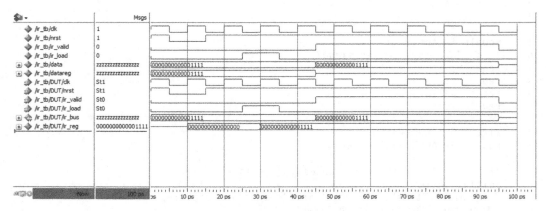

Figure 8.6
Basic processor instruction register simulation.

separate input and output setting for the memory; therefore, the MDR becomes a simple register which sets its output to the value of the memory output when its control signal mdr_load is high. The Memory Data Register (MDR) in a simple microprocessor needs enough bits for the address and opcode combined. For example, if the address requires 8 bits, and the opcode also requires 8 bits, then the size of the register needs to be 16 bits wide (8 + 8). If the output from the Memory Data Register goes onto the main bus, then this can be read into the instruction register, which is also 16 bits wide in this case.

The Memory Data Register (MDR) therefore has clock and reset signals, and also the same interface to the internal processor bus (mdr_bus) defined as a standard logic of direction inout. The MDR also has a further control signal, to make the required data available on the system bus (mdr_valid). This consists of the opcode and address, which can be used by the controller, Accumulator or Instruction register.

The code for the Memory Data Register (MDR) is therefore given in the following listing:

```
1     'define OP 8
2     'define ADDR 8
3
4     module mdr (clk,nrst,mdr_load,mdr_valid,mem_bus,mdr_bus);
5
6     input clk;
7     input nrst;
8     input mdr_valid;
9     input mdr_load;
10    input mem_bus;
11
12    inout ['OP+'ADDR-1:0] mdr_bus;
13    input ['OP+'ADDR-1:0] mem_bus;
14
15    wire ['OP+'ADDR-1:0] mdr_bus;
16    wire ['OP+'ADDR-1:0] mem_bus;
17
18    reg ['OP+'ADDR-1:0] mdr_reg;
19
20    assign mdr_bus = mdr_valid ? mdr_reg : 16'bz;
21
22    always @(posedge clk) begin
23      if(nrst==0) begin
24        mdr_reg <= 0;
25      end
26      else begin
27        if(mdr_load==1) begin
28          mdr_reg <= mem_bus;
29        end
30      end
31    end
32
33    endmodule
```

We can test this by loading in a sample instruction, and then setting it valid so that it is seen on the bus. The test bench to achieve this is shown here:

```
1    `define OP 8
2    `define ADDR 8
3
4
5    module mdr_tb();
6    // declare the counter signals
7    reg clk;
8    reg nrst;
9    reg mdr_valid;
10   reg mdr_load;
11
12   wire ['OP+'ADDR—1:0] data;
13   reg ['OP+'ADDR—1:0] memory;
14
15   // Set up the initial variables and reset
16   initial begin
17     $display ("time\t clk reset inc load valid data");
18     $monitor ("%g\t %b %b %b %b %b",
19     $time, clk, nrst, mdr_load, mdr_valid, data);
20     clk = 1;    // initialize the clock to 1
21     nrst = 1;   // set the reset to 1 (not reset)
22     mdr_valid = 0;
23     mdr_load = 0;
24     memory = 16'b0000000000001111;
25     #5 nrst = 0; // reset = 0 : resets the counter
26     #10 nrst = 1; // reset back to 1 : counter can start
27     #10 mdr_load = 1;
28     #10 mdr_load = 0;
29     #10 memory = 16'bzzzzzzzzzzzzzzzz;
30     mdr_valid = 1;
31     #50 mdr_valid = 0; // reset back to 1 : counter can start
32     #200 $finish; // Finish the simulation
33   end
34
35   // Clock generator
36   always begin
37     #5 clk = ~clk; // Clock every 5 time slots
38   end
39
40   //assign data = datareg;
41
42   // Connect DUT to test bench
43   mdr DUT ( clk,nrst,mdr_load,mdr_valid,memory,data);
44
45   endmodule
```

The resulting waveform shows the behavior as predicted (Figure 8.7):

8.4.4 Memory Address Register

The memory address register is used to handle the address transferred to the memory unit, and this can be handled either using a bus approach (which we have used in this architecture) or

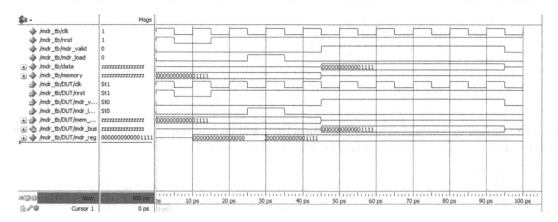

Figure 8.7
Basic processor memory data register (MDR) simulation.

direct input declaration for the memory. In this case we will use a bus setting for the memory, therefore the MAR becomes a simple register which sets its output to the value of the required address from the IR or PC when its control signal mar_load is high. The Memory Address Register (MAR) in a simple microprocessor needs enough bits for the address. For example, if the address requires 8 bits then the The size of the register needs to be 8 bits wide.

The Memory Address Register (MAR) therefore has clock and reset signals, and also the same interface to the internal processor bus (mar_bus) defined as a standard logic of direction inout, however only the first 8 bits are used.

The code for the Memory Address Register (MAR) is therefore given in the listing below

```
1    'define ADDR 8
2    'define OP 8
3
4    module mar (clk,nrst,mar_load,mar_bus,address);
5
6    input clk;
7    input nrst;
8    input mar_load;
9
10   input ['OP+'ADDR-1:0] mar_bus;
11   output ['ADDR-1:0] address;
12
13   wire ['OP+'ADDR-1:0] mar_bus;
14   reg ['ADDR-1:0] address;
15
16   always @(posedge clk) begin
17     if(nrst==0) begin
18       address <= 0;
19     end
20     else begin
```

```
21          if(mar_load==1) begin
22            address <= mar_bus['ADDR-1:0];
23          end
24        end
25      end
26
27      endmodule
```

We can test this by loading in a sample instruction, and then setting it valid so that it is seen on the bus. The test bench to achieve this is shown below:

```
1     'define OP 8
2     'define ADDR 8
3
4
5     module mar_tb();
6     // declare the counter signals
7     reg clk;
8     reg nrst;
9     reg mar_load;
10
11    reg ['OP+'ADDR-1:0] data;
12    wire ['ADDR-1:0] address;
13
14    // Set up the initial variables and reset
15    initial begin
16      $display ("time\t clk reset inc load valid data");
17      $monitor ("%g\t %b %b %b %b %b",
18      $time, clk, nrst, mar_load, data, address);
19      clk = 1;    // initialize the clock to 1
20      nrst = 1;    // set the reset to 1 (not reset)
21      mar_load = 0;
22      data = 16'b0000000000001111;
23      #5 nrst = 0; // reset = 0 : resets the counter
24      #10 nrst = 1; // reset back to 1 : counter can start
25      #10 mar_load = 1;
26      #10 mar_load = 0;
27      #10 data = 16'bzzzzzzzzzzzzzzzz;
28      #200 $finish; // Finish the simulation
29    end
30
31    // Clock generator
32    always begin
33      #5 clk = ~clk; // Clock every 5 time slots
34    end
35
36    //assign data = datareg;
37
38    // Connect DUT to test bench
39    mar DUT ( clk,nrst,mar_load,data,address);
40
41    endmodule
```

The resulting waveform shows the behavior as predicted (Figure 8.8):

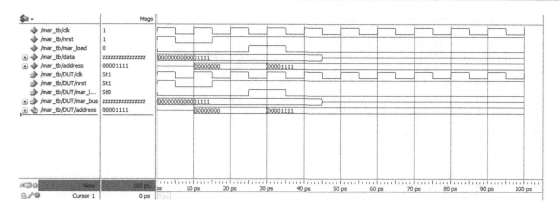

Figure 8.8
Basic processor memory address register (MAR) simulation.

8.4.5 The Arithmetic and Logic Unit

The Arithmetic and Logic Unit (ALU) has the same clock and reset signals as the PC, and also the same interface to the bus (alu_bus) defined as a type inout. The ALU also has three further control signals, which can be decoded to map to the 8 individual functions required of the ALU. The ALU also contains the Accumulator (ACC) which is an input of the size defined for the system bus width. There is also a single bit output alu_zero which goes high when all the bits in the accumulator are zero.

The function of the ALU is to decode the alu_op in binary form and then carry out the relevant function on the data on the bus, and the current data in the accumulator. If the alu_valid is low, the the bus value should be set to Z for all bits. If the reset signal (nest) is low, then the register value internally should be set to all 0. On the rising edge of the clock, the value on the bus shall be sent to the internal register and the command shall be decoded. The resulting Verilog model is given as follows:

```
1    'define OP 8
2    'define ADDR 8
3
4    module alu (clk, nrst, alu_op, alu_zero, alu_valid, alu_bus);
5
6    // Interface Definitions
7    input clk;   // Clock Input
8    input nrst ;   // reset (active Low)
9    output alu_zero; // ALU is zero
10   input alu_valid; // ALU output is valid
11
12   inout ['OP+'ADDR-1:0] alu_bus; // ALU bus
13   input ['OP-1:0] alu_op;   // ALU OP code
14
15   // Register Definitions
```

```
16    reg alu_zero; // ALU is zero
17
18    reg ['OP+'ADDR−1:0] acc; // Accumulator
19    reg ['OP+'ADDR−1:0] alu_reg; // Accumulator Reg
20
21    assign alu_bus = alu_valid ? alu_reg : 'bz;
22
23    always @(posedge clk) begin
24
25      if(nrst==0) begin
26        acc <= 0;
27      end
28      else begin
29        case (alu_op)
30          8'h00: acc <= alu_bus;
31          8'h01: acc <= acc + alu_bus;
32          8'h02: acc <= ~alu_bus;
33          8'h03: acc <= acc | alu_bus;
34          8'h04: acc <= acc & alu_bus;
35          8'h05: acc <= acc ^ alu_bus;
36          8'h06: acc <= acc + 1;
37          8'h07: alu_reg <= acc;
38          default: acc = 0;
39        endcase
40        if (acc==0) begin
41          alu_zero <=1;
42        end
43        else begin
44          alu_zero <= 0;
45        end
46      end
47    end
48
49    endmodule
```

We can test this by loading in a sample instruction, and then setting it valid so that it is then seen on the bus. The test bench to achieve this is shown here, and in this case after initializing the accumulator to all zeros, the hex value 0012 is loaded, with the binary equivalent 0000 0000 0001 0010, which is seen in Figure 8.9.

```
1     'define OP 8
2     'define ADDR 8
3
4
5     module alu_tb();
6     // declare the counter signals
7     reg clk;
8     reg nrst;
9     reg alu_valid;
10    wire alu_zero;
11
12    wire ['OP+'ADDR−1:0] alu_bus;
13    reg ['OP−1:0] opcode;
```

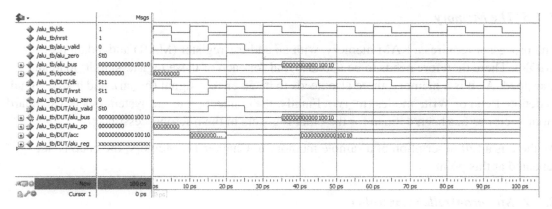

Figure 8.9
Basic processor ALU simulation.

```
14    reg ['OP+'ADDR-1:0] alu_reg;
15
16
17    assign alu_bus = alu_reg;
18
19    // Set up the initial variables and reset
20    initial begin
21      $display ("time\t clk reset inc load valid data");
22      $monitor ("%g\t %b %b %b %b %b",
23      $time, clk, nrst, alu_zero, alu_valid, opcode, alu_bus);
24      clk = 1;     // initialize the clock to 1
25      nrst = 1;          // set the reset to 1 (not reset)
26      alu_valid = 0;
27      opcode = 8'h00;
28      alu_reg = 'bz;
29      #5 nrst = 0;     // reset = 0 : resets the counter
30      #10 nrst = 1;     // reset back to 1 : counter can start
31      alu_valid = 1;
32      #10 alu_valid = 0;   // reset back to 1 : counter can start
33      #10 alu_reg = 16'h0012;
34      #200 $finish; // Finish the simulation
35    end
36
37    // Clock generator
38    always begin
39      #5 clk = ~clk;   // Clock every 5 time slots
40    end
41
42    //assign data = datareg;
43
44    // Connect DUT to test bench
45    alu DUT ( clk,nrst,opcode,alu_zero,alu_valid,alu_bus);
46
47    endmodule
```

The resulting waveform shows the behavior as predicted:

8.4.6 The Memory

The processor requires a RAM memory, with an address register (MAR) and a data register (MDR). There therefore needs to be a load signal for each of these registers: mdr_load and mar_load. As it is a memory, there also needs to be an enable signal (m_en), and also a signal to denote Read or Write modes (m_rw). Finally, the connection to the system bus is a standard inout vector as has been defined for the other registers in the microprocessor.

As there is a full description of a sample memory in Chapter 11, Memory, the code is not repeated at this point.

8.4.7 Microcontroller Controller

The operation of the processor is controlled in detail by the sequencer, or controller block. The function of this part of the processor is to take the current program counter address, look up the relevant instruction from memory, move the data around as required, setting up all the relevant control signals at the right time, with the right values. As a result, the controller must have the clock and reset signals (as for the other blocks in the design), a connection to the global bus, and finally all the relevant control signals must be output.

Using this entity, the control signals for each separate block are then defined, and these can be used to carry out the functionality requested by the program. The architecture for the controller is then defined as a basic state machine to drive the correct signals. The basic state machine for the processor is defined in Figure 8.4 shown previously in this chapter.

The outline Verilog Controller is shown below; however, there are so many states it has been cut down to illustrate the architecture of the model.

```
1    `define OP 8
2    `define ADDR 8
3
4    module controller (
5      clk, nrst,
6      ir_load, ir_valid, ir_address,
7      pc_inc,pc_load, pc_valid,
8      mdr_load,
9      mar_load, mar_valid,
10     m_en, m_rw,
11     alu_op,
12     alu_valid
13   );
14
15   // Interface Definitions
16   input clk;   // Clock Input
17   input nrst ;  // reset (active Low)
18
19   output ir_load;
20   output ir_valid;
```

```
21      input ['ADDR-1:0] ir_address;
22
23      output pc_inc;
24      output pc_load;
25      output pc_valid
26
27      output alu_valid; // ALU output is valid
28
29      output mdr_load;
30      output mar_load;
31      output mar_valid;
32      output m_en;
33      output m_rw;
34
35      output ['OP-1:0] alu_op;  // ALU OP code
36
37      // Register Definitions
38
39      reg ir_load;
40      reg ir_valid;
41
42      reg pc_inc;
43      reg pc_load;
44      reg pc_valid
45
46      reg alu_valid;   // ALU output is valid
47
48      reg mdr_load;
49      reg mar_load;
50      reg mar_valid;
51      reg m_en;
52      reg m_rw;
53
54      reg ['OP-1:0] alu_op; // ALU OP code
55
56      reg [3:0] state; // state variable
57
58      always @(posedge clk) begin
59
60        if(nrst==0) begin
61          acc <= 0;
62        end
63        else begin
64          case (state)
65            8'h00:  begin
66                mar_load <= 1;
67                pc_load <= 1;
68                pc_inc <= 1;
69              end
70
71          // Complete State Definitions Here
72
73
74          // Catch All state to avoid unknown conditions
75          default: state <= 0;
```

```
76        endcase
77        if (acc==0) begin
78          alu_zero <=1;
79        end
80        else begin
81          alu_zero <= 0;
82        end
83      end
84    end
85
86    always @(posedge clk or posedge rst)
87    begin
88      if (rst == 0)
89        state = s0;
90      else
91        case (state)
92          s0:
93            state = s1;
94          s1:
95            if (choice)
96              state = s3;
97            else
98              state = s2;
99          s2:
100            state = s0;
101          s3:
102            state = s0;
103        endcase
104    end
105
106    endmodule
```

8.4.8 Summary of a Simple Verilog Microprocessor

Now that the important elements of the processor have been defined, it is a simple matter to instantiate them in a complete Verilog model and create a microprocessor using these building blocks. It is also a simple matter to modify the functionality of the processor by changing the address/data bus widths or extend the instruction set.

8.5 Soft Core Processors on an FPGA

While the previous example of a simple microprocessor is useful as a design exercise and helpful to gain understanding about how microprocessors operate, in practice most FPGA vendors provide standard processor cores as part of an embedded development kit that includes compilers and other libraries. For example, this could be the MicroBlaze™ core from Xilinx or the Nios™ core supplied by Altera. In all these cases the basic idea is the same: that a standard configurable core can be instantiated in the design and code compiled using a standard compiler and downloaded to the processor core in question.

Each soft core is different and rather than describe the details of a particular case, in this section the general principles will be covered and the reader is encouraged to experiment with the offerings from the FPGA vendors to see which suits their application the best.

In any soft core development system there are several key functions that are required to make the process easy to implement. The first is the system building function. This enables a core to be designed into a hardware system that includes memory modules, control functions, DMA functions, data interfaces, and interrupts. The second is the choice of processor types to implement. A basic Nios II or similar embedded core will typically have a performance in the region of 100-200MIPS, and the processor design tools will allow the size of the core to be traded off with the hardware resources available and the performance required.

8.6 Summary

The topic of embedded processors on FPGAs would be suitable for a complete book in itself. In this chapter the basic techniques have been described for implementing a simple processor directly on the FPGA and the approach for implementing soft cores on FPGAs have been introduced.

PART 3

Designer's Toolbox

This part of the book is intended to provide the designer with some useful building blocks that could be used in a typical FPGA-based system. These are not production quality code by any means, but rather a short cut to aid understanding and help designers make good choices in their own integration of functions into a design.

The chapters in this part include useful techniques such as serial communications, secure communications, memory, handling peripherals and digital filters. It is a snapshot, and by its nature a subset of what is possible. However, hopefully it will be of assistance to designers in their selection of functionality to include in their own designs.

CHAPTER 9

Digital Filters

9.1 Introduction

An important part of digital or computing systems that interface to the "real/world" of sensors and analog interfaces is the ability to process sampled data in the digital domain. This is often called Sampled Data Systems (SDS) or defined as operating in the Z-domain. Most engineers are familiar with the operation of filters in the Laplace or S-domain where a continuous function defines the characteristics of the filter and this is the digital domain equivalent to that.

For example, consider a simple RC circuit in the analog domain, which is designed to be a low pass filter, as shown in Figure 9.1.

This has a low pass filter behavior and can be represented mathematically using the continuous Laplace (or S-domain) notation:

$$L(s) = \frac{1}{1 + sRC} \tag{9.1}$$

This function is a low pass filter because the Laplace operator s is equivalent to $j\omega$, where $\omega = 2\pi f$ (with f being the frequency). If f is zero (the d.c. condition), then the gain will be 1, but if the value of sRC is equal to 1, then the gain will be 0.5. This in dB is -3 dB and is the classical low pass filter cut-off frequency.

In the digital domain, the s operation is replaced by Z. Z^{-1} is equivalent in a practical sense to a delay operator, and similar functions to the Laplace filter equations can be constructed for the digital, or Z domain, equivalent.

There are a number of design techniques, many beyond the scope of this book (if the reader requires a more detailed introduction to the realm of digital filters, Cunningham's *Digital Filtering: An Introduction* is a useful starting point); however, it is useful to introduce some of the basic techniques used in practice and illustrate them with examples.

The remainder of this chapter will cover the introduction to the basic techniques and then demonstrate how these can be implemented using VHDL and Verilog on FPGAs.

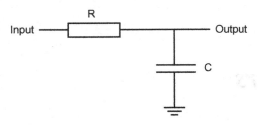

Figure 9.1
RC filter in the analog domain.

9.2 Converting S Domain to Z Domain

The method of converting an S domain equation for a filter to its equivalent Z domain expression uses the *bilinear transform*. This is a standard method for expressing the S-domain equation in the Z-domain. The basic approach is to replace each instance of *s* with its equivalent Z domain notation and then rearrange into the most convenient form. The transform is called bilinear as both the numerator and denominator of the expression are linear in terms of *z*.

$$s = \frac{z-1}{z+1} \tag{9.2}$$

If we take a simple example of a basic second order filter we can show how this is translated into the equivalent Z domain form:

$$H(s) = \frac{1}{s^2 + 2s + 1} \tag{9.3}$$

In order to get the function $H(s)$ in its Z-domain equivalent, replace each occurrence of *s* with the expression for *s* in terms of *z* shown in Equation (9.2), giving:

$$H(z) = \frac{1}{(\frac{z-1}{z+1})^2 + 2(\frac{z-1}{z+1}) + 1} \tag{9.4}$$

$$H(z) = \frac{(z+1)^2}{(z-1)^2 + (z-1)(z+1) + (z+1)^2} \tag{9.5}$$

$$H(z) = \frac{z^2 + 2z + 1}{3z^2 + 1} \tag{9.6}$$

Now, the term $H(z)$ is really the output $Y(z)$ over the input $X(z)$ and we can use this to express the Z domain equation in terms of the input and output:

$$H(z) = \frac{z^2 + 2z + 1}{3z^2 + 1} = \frac{Y(z)}{X(z)} \tag{9.7}$$

This can then be turned into a sequence expression using delays (z is one delay, z^2 is two delays and so on) with the following result:

$$3z^2Y(z) + Y(z) = z^2X(z) + 2zX(z) + X(z) \tag{9.8}$$

$$3y(n + 2) + y(n) = x(n + 1) + 2x(n + 1) + x(n) \tag{9.9}$$

This is useful because we are now expressing the Z domain equation in terms of delay terms, and the final step is to express the value of $y(n)$ (the current output) in terms of past elements by reducing the delays accordingly (by 2 in this case):

$$3y(n) + y(n - 2) = x(n) + 2x(n - 1) + x(n - 2) \tag{9.10}$$

$$y(n) + 1/3y(n - 2) = 1/3x(n) + 2/3x(n - 1) + 1/3x(n - 2) \tag{9.11}$$

$$y(n) = 1/3x(n) + 2/3x(n - 1) + 1/3x(n - 1) - 1/3y(n - 2) \tag{9.12}$$

The final design note at this point is to make sure that the design frequency is correct, for example the low pass cut-off frequency. The frequencies are different between the S and Z domain models, even after the bilinear transformation, and in fact the desired digital domain frequency must be translated into the equivalent s domain frequency using a technique called prewarping. This simple step translates the frequency from one domain to the other using the following expression:

$$\omega_c = \tan\left(\frac{\Omega_c T}{2}\right) \tag{9.13}$$

where Ω_c is the digital domain frequency, T is the sampling period of the Z domain system and ω_c is the resulting frequency for the analog domain calculations.

Once we have obtained our Z domain expressions, how do we turn this into practical designs? The next section will explain how this can be achieved in VHDL.

9.3 Implementing Z Domain Functions in VHDL

9.3.1 Introduction

Z domain functions are essentially digital in the time domain as they are discrete and sampled. The functions are also discrete in the amplitude axis, as the variables or signals are defined using a fixed number of bits in a real hardware system; whether this is integer, signed, fixed point, or floating point, there is always a finite resolution to the signals. For the remainder of this chapter, signed arithmetic is assumed for simplicity and ease of understanding. This also essentially defines the number of bits to be used in the system. If we have 8 bits, the resolution is 1 bit and the range is -128 to $+127$.

9.3.2 Gain Block

The first main Z domain block is a simple gain block. This requires a single signed input, a single signed output and a parameter for the gain. This could be an integer or also a signed value. The VHDL model for a simple Z domain gain block is given as:

```
1    library ieee;
2    use ieee.numeric_std.all;
3
4    entity zgain is
5      generic ( n : integer := 8;
6        gain : signed
7        );
8      port (
9        zin : in signed ( n-1 downto 0 );
10       zout : out signed (n-1 downto 0)
11       );
12   end entity zgain;
13
14   architecture zdomain of zgain is
15   begin
16     p1 : process(zin)
17       variable product : signed ( 2*n-1 downto 0);
18     begin
19       product := zin * gain;
20       zout <= product (n-1 downto 0 );
21     end process p1;
22   end architecture zdomain;
```

We can test this with a simple testbench that ramps up the input and we can observe the output being changed in turn:

```
1    library ieee;
2    use ieee.std_logic_1164.all;
3    use ieee.numeric_std.all;
4
5    entity tb is
6    end entity tb;
7
8    architecture testbench of tb is
9
10     signal clk : std_logic := '0';
11     signal dir : std_logic := '0';
12     signal zin : signed (7 downto 0):=X"00";
13     signal zout : signed (7 downto 0):=X"00";
14
```

```
15      component zgain
16      generic (
17        n : integer := 8;
18        gain :signed := X"02"
19        );
20      port (
21        signal zin : in signed(n-1 downto 0);
22        signal zout : out signed(n-1 downto 0)
23        );
24      end component;
25      for all : zgain use entity work.zgain;
26
27
28    begin
29      clk <= not clk after 1 us;
30
31      DUT : zgain generic map ( 8, X"02" ) port map ( zin, zout);
32
33      p1 : process (clk)
34      begin
35        zin <= zin + 1;
36      end process p1;
37    end architecture testbench;
```

Clearly, this model has no error checking or range checking and the obvious problem with this type of approach is that of overflow. For example, if we multiply the input (64) by a gain of 2, we will get 128, but that is the sign bit, and so the result will show −128! This is an obvious problem with this simplistic model and care must be taken to ensure that adequate checking takes place in the model.

9.3.3 Sum and Difference

Using this same basic approach, we can create sum and difference models which are also essential building blocks for a Z domain system. The sum model VHDL is shown here:

```
1     library ieee;
2     use ieee.numeric_std.all;
3
4     entity zsum is
5       generic ( n : integer := 8
6         );
7       port (
8         zin1 : in signed ( n-1 downto 0 );
9         zin2 : in signed ( n-1 downto 0 );
10        zout : out signed (n-1 downto 0)
11        );
12    end entity zsum;
13
14    architecture zdomain of zsum is
15    begin
```

```
16        p1 : process(zin)
17          variable zsum : signed ( 2*n-1 downto 0);
18        begin
19          zsum := zin1 + zin2;
20          zout <= zsum (n-1 downto 0);
21        end process p1;
22      end architecture zdomain;
```

Despite the potential for problems with overflow, both of the models shown have the internal variable that is twice the number of bits required, and so this can take care of any possible overflow internal to the model, and in fact checking could take place prior to the final assignment of the output to ensure the data is correct. The difference model is almost identical to the sum model except that the difference of zin1 and zin2 is computed.

9.3.4 Division Model

A useful model for scaling numbers simply in the Z domain is the division by 2 model. This model simply shifts the current value in the input to the right by one bit, hence giving a division by 2. The model could easily be extended to shift right by any number of bits, but this simple version is very useful by itself. The VHDL for the model relies on the logical shift right operator (SRL) which not only shifts the bits right (losing the least significant bit) but adds a zero at the most significant bit. The resulting VHDL is shown for this specific function:

```
1        zout <= zin srl 1;
```

The unit shift can be replaced by any integer number to give a shift of a specific number of bits. For example, to shift right by 3 bits (effectively a divide by 8) would have the following VHDL:

```
1        zout <= zin srl 3;
```

The complete division by 2 model is given here:

```
1      library ieee;
2      use ieee.numeric_std.all;
3
4      entity zdiv2 is
5        generic ( n : integer := 8
6          );
7        port (
8          zin : in signed ( n-1 downto 0 );
9          zout : out signed (n-1 downto 0)
10          );
11      end entity zdiv2;
12
13      architecture zdomain of zdiv2 is
14      begin
```

```
15        zout <= zin srl 1;
16      end architecture zdomain;
```

In order to test the model a simple test circuit that ramps up the input is used and this is given as follows:

```
1      library ieee;
2      use ieee.numeric_std.all;
3
4      entity zdiv2 is
5        generic ( n : integer := 8
6          );
7        port (
8          zin : in signed ( n—1 downto 0 );
9          zout : out signed (n—1 downto 0)
10         );
11     end entity zdiv2;
12
13     architecture zdomain of zdiv2 is
14     begin
15       zout <= zin srl 1;
16     end architecture zdomain;
```

The behavior of the model is useful to review. If the input is X03 (Decimal 3), binary 00000011 and the number is right shifted by one, then the resulting binary number will be 00000001 (X01 or decimal 1); in other words this operation always rounds down. This has obvious implications for potential loss of accuracy and the operation is skewed downward, which has, again, implications for how numbers will be treated using this operator in a more complex circuit.

9.3.5 Unit Delay Model

The final basic model is the unit delay model (zdelay). This has a clock input (clk) using a std_logic signal to make it simple to interface to standard digital controls. The output is simply a one clock cycle delayed version of the input.

Notice that the output zout is initialized to all zeros for the initial state; otherwise don't care conditions can result that propagate across the complete model.

```
1      library ieee;
2      use ieee.std_logic_1164.all;
3      use ieee.numeric_std.all;
4
5      entity zdelay is
6        generic ( n : integer := 8 );
7        port (
8          clk : in std_logic;
9          zin : in signed ( n—1 downto 0 );
10         zout : out signed (n—1 downto 0) := (others => 0 )
11         );
```

```
12    end entity zdelay;
13
14    architecture zdomain of zdelay is
15      signal lastzin : signed (n-1 downto 0) := (others => '0');
16    begin
17      p1 : process(clk)
18      begin
19        if rising_edge(clk) then
20          zout <= lastzin;
21          lastzin <= zin;
22        end if;
23      end process p1;
24    end architecture zdomain;
```

9.4 Basic Low Pass Filter Model

We can put these elements together in simple models that implement basic filter blocks in any configuration we require, as always taking care to ensure that overflow errors are checked for in practice.

To demonstrate this, we can implement a simple low pass filter using the basic block diagram shown in Figure 9.2.

We can create a simple test circuit that uses the individual models we have already shown for the sum and delay blocks and apply a step change and observe the response of the filter to this stimulus. Clearly, in this case, with unity gain the filter exhibits positive feedback and so to ensure the correct behavior we use the divide by 2 model zdiv2 in both the inputs to the sum block to ensure gain of 0.5 on both. These are not shown in the figure. The resulting VHDL model is shown in the following code (note the use of the zdiv2 model):

```
1    library ieee;
2    use ieee.std_logic_1164.all;
3    use ieee.numeric_std.all;
4
```

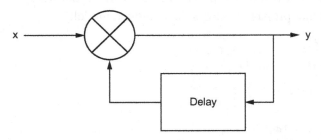

Figure 9.2
Simple Z domain low pass filter.

```
 5    entity tb is
 6    end entity tb;
 7
 8    architecture testbench of tb is
 9
10      signal clk : std_logic := '0';
11      signal x : signed (7 downto 0):=X"00";
12      signal y : signed (7 downto 0):=X"00";
13      signal y1 : signed (7 downto 0):=X"00";
14      signal yd : signed (7 downto 0):=X"00";
15      signal yd2 : signed (7 downto 0):=X"00";
16      signal x2 : signed (7 downto 0):=X"00";
17
18      component zsum
19      generic (
20        n : integer := 8
21      );
22      port (
23        signal zin1 : in signed(n-1 downto 0);
24        signal zin2 : in signed(n-1 downto 0);
25        signal zout : out signed(n-1 downto 0)
26      );
27      end component;
28      for all : zsum use entity work.zsum;
29
30      component zdiff
31      generic (
32        n : integer := 8
33      );
34      port (
35        signal zin1 : in signed(n-1 downto 0);
36        signal zin2 : in signed(n-1 downto 0);
37        signal zout : out signed(n-1 downto 0)
38      );
39      end component;
40      for all : zdiff use entity work.zdiff;
41
42      component zdiv2
43      generic (
44        n : integer := 8
45      );
46      port (
47        signal zin : in signed(n-1 downto 0);
48        signal zout : out signed(n-1 downto 0)
49      );
50      end component;
51      for all : zdiv2 use entity work.zdiv2;
52
53      component zdelay
54      generic (
55        n : integer := 8
56      );
57      port (
58        signal clk : in std_logic;
59        signal zin : in signed(n-1 downto 0);
```

```
60        signal zout : out signed(n−1 downto 0)
61      );
62    end component;
63    for all : zdelay use entity work.zdelay;
64
65    begin
66      clk <= not clk after 1 us;
67
68      GAIN1 : zdiv2 generic map (8) port map ( x, x2);
69      GAIN2 : zdiv2 generic map (8) port map ( yd, yd2);
70      SUM1 : zsum generic map ( 8 ) port map ( x2, yd2, y );
71      D1 : zdelay generic map ( 8 ) port map ( clk, y, yd );
72
73      x <= X"00", X"0F" after 10 us;
74    end architecture testbench;
```

The test circuit applies a step change of X00 to X0F after 10 μs, and this results in the filter response. We can show this graphically in Figure 9.3 with the output in both hexadecimal and analog form for illustration.

It is interesting to note the effect of using the zdiv2 function on the results. With the input of 0F (binary 00001111) we lose the LSB when we divide by 2, giving the resulting input to the sum block of 00000111 (7) which added together with the division of the output gives a total of 14 as the maximum possible output from the filter. In fact, the filter gives an output of X0D or binary 00001101, which is two down from the theoretical maximum of X0F and this highlights the practical difficulties when using a coarse approximation technique for numerical work rather than a fixed or floating point method. On the other hand, it is clearly a simple and effective method of implementing a basic filter in VHDL.

Later in this book, the use of fixed and floating point numbers are discussed, as is the use of multiplication for more exact calculations and for practical filter design; where higher

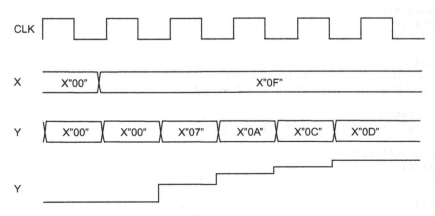

Figure 9.3
Basic low pass filter simulation waveforms.

accuracy is required, then it is likely that both these methods would be used. There may be situations, however, where it is simply not possible to use these advanced techniques, particularly a problem when space is at a premium on the FPGA and, in these cases, the simple approach described in this chapter will be required.

There are numerous texts on more advanced topics in digital filter design, and these are beyond the scope of this book, but it is useful to introduce some key concepts at this stage of the two main types of digital filter in common usage today. These are the recursive (or Infinite Impulse Response, IIR) filters and nonrecursive (or Finite Impulse Response, FIR) filters.

9.5 Implementing Z Domain Functions in Verilog

9.5.1 Gain Block

The first main Z domain block is a simple gain block. This requires a single signed input, a single signed output and a parameter for the gain. This could be an integer or also a signed value. The Verilog model for a simple Z domain gain block is given as follows:

```
1   module zgain (
2     din, // Digital Input
3     dout // Digital Output
4     );
5
6   parameter n = 8; // Width of Digital Input and Output
7   parameter gain = 1; // Gain Parameter
8
9   input [n-1:0] din;
10  output [2*n-1:0] dout;
11
12  wire signed [n-1:0] din;
13  reg signed [2*n-1:0] dout;
14
15  always @ (din)
16  begin
17    dout <= din * gain;
18  end
19
20  endmodule
```

Clearly, as with the VHDL model, this model has no error checking or range checking and the obvious problem with this type of approach is that of overflow. For example, in an 8-bit model, if we multiply the input (64) by a gain of 2, we will get 128, but that is the sign bit, and so the result will show −127! This is an obvious problem with this simplistic model and care must be taken to ensure that adequate checking takes place in the model.

We can test the model by using a simple test bench that has a lookup table to generate a sine wave which will go from −127 to +127, and the gain can be varied using the parameter of the gain block. The number of bits in the gain block defaults to the parameter *n* value (default is 8) and the output is $2*n$. This makes the assumption that the gain will be no larger than the input value maximum, and this should obviously be checked thoroughly in a practical model to avoid the possibility of overflow errors occurring. The test bench Verilog is shown here:

```
1   module zsine_tb();
2   // declare the counter signals
3   reg clk;
4   reg signed [7:0] zvalue;
5   reg rst;
6
7   wire signed [15:0] gvalue;
8   // Set up the initial variables and reset
9   initial begin
10    $display ("time\t clk zvalue");
11    $monitor ("%g\t %b %d",
12    $time, clk, zvalue);
13    clk = 1;        // initialize the clock to 1
14    rst = 1;        // set the reset to 1 (not reset)
15    #5 rst = 0;     // reset = 0 : resets the counter
16    #10 rst = 1;    // reset back to 1 : counter can start
17    #5 zvalue = 0;
18    #5 zvalue = 22;
19    #5 zvalue = 44;
20    #5 zvalue = 64;
21    #5 zvalue = 82;
22    #5 zvalue = 98;
23    #5 zvalue = 111;
24    #5 zvalue = 120;
25    #5 zvalue = 126;
26    #5 zvalue = 127;
27    #5 zvalue = 126;
28    #5 zvalue = 120;
29    #5 zvalue = 111;
30    #5 zvalue = 98;
31    #5 zvalue = 82;
32    #5 zvalue = 64;
33    #5 zvalue = 44;
34    #5 zvalue = 22;
35    #5 zvalue = 0;
36    #5 zvalue = -22;
37    #5 zvalue = -44;
38    #5 zvalue = -64;
39    #5 zvalue = -82;
40    #5 zvalue = -98;
41    #5 zvalue = -111;
42    #5 zvalue = -120;
43    #5 zvalue = -126;
44    #5 zvalue = -127;
45    #5 zvalue = -126;
46    #5 zvalue = -120;
47    #5 zvalue = -111;
48    #5 zvalue = -98;
49    #5 zvalue = -82;
50    #5 zvalue = -64;
51    #5 zvalue = -44;
```

```
52        #5 zvalue = -22;
53        #5 zvalue = 0;
54        #1000 $finish;    // Finish the simulation
55      end
56
57      // Clock generator
58      always begin
59        #5 clk = ~clk; // Clock every 5 time slots
60      end
61
62      zgain #(8,2) a1(zvalue, gvalue);
63
64      endmodule
```

When the model was simulated with a gain of 2, $n = 8$, the output values can be seen in Figure 9.4 to exactly double the input values, and as the gain block sensitivity is on the input, the change is immediate (unlike a synchronous model which would only change on the clock edge). The waveforms are shown in Figure 9.4.

When the gain is reversed to -2, the output has the same amplitude as before, but now the output is inverted, and this can easily be seen to be correct in the waveform diagram in Figure 9.5.

9.5.2 Sum and Difference

Using this same basic approach, we can create sum and difference models which are also essential building blocks for a Z domain system. The sum model Verilog is shown here:

```
1       module zsum (
2         din1, // Digital Input 1
3         din2, // Digital Input 2
4         dout // Digital Output
5         );
6
7       parameter n = 8; // Width of Digital Input and Output
```

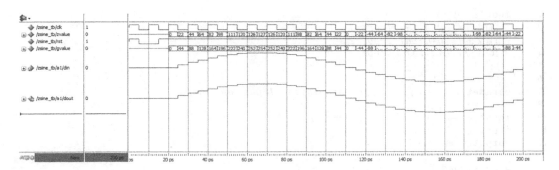

Figure 9.4
Z gain simulation test with gain = 2.

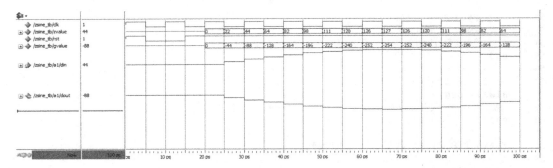

Figure 9.5
Z gain simulation test with gain $= -2$.

```
8
9    input [n−1:0] din1;
10   input [n−1:0] din2;
11   output [2*n−1:0] dout;
12
13   wire signed [n−1:0] din1;
14   wire signed [n−1:0] din2;
15   reg signed [2*n−1:0] dout;
16
17   always @ (din1 or din2)
18   begin
19     dout <= din1 + din2;
20   end
21
22   endmodule
```

Despite the potential for problems with overflow, both of the models shown have the internal variable that is twice the number of bits required, and so this can take care of any possible overflow internal to the model, and in fact checking could take place prior to the final assignment of the output to ensure the data is correct. The difference model is almost identical to the sum model except that the difference of din1 and din2 is computed.

```
1    module zdiff (
2      din1, // Digital Input 1
3      din2, // Digital Input 2
4      dout // Digital Output
5      );
6
7    parameter n = 8; // Width of Digital Input and Output
8
9    input [n−1:0] din1;
10   input [n−1:0] din2;
11   output [2*n−1:0] dout;
12
13   wire signed [n−1:0] din1;
14   wire signed [n−1:0] din2;
15   reg signed [2*n−1:0] dout;
```

```
16
17    always @ (din1 or din2)
18    begin
19      dout <= din1 — din2;
20    end
21
22    endmodule
```

9.5.3 Unit Delay Model

The final basic model is the unit delay model (zdelay). This has a clock input (clk) and a reset (rst) signal to make it simple to interface to standard digital controls. The output is simply a one clock cycle delayed version of the input.

Notice that the output dout is initialized to all zeros for the initial state, otherwise *don't care* conditions can result that propagate across the complete model.

```
1     module zdelay (
2       clk, // clock input
3       rst, // reset input
4       din, // Digital Input
5       dout // Digital Output
6       );
7
8     parameter n = 8; // Width of Digital Input and Output
9
10    input clk;
11    input rst;
12    input [n—1:0] din;
13    output [n—1:0] dout;
14
15    reg [n—1:0] dout;
16
17    reg [n—1:0] dstored;
18
19    always @ (posedge clk)
20    begin
21      if (rst == 0 ) begin
22        dout <= 0;
23        dstored <= 0;
24      end
25      else begin
26        dout <= dstored;
27        dstored <= din;
28      end
29    end
30
31    endmodule
```

The model was tested using a basic test bench as shown below to illustrate how the simple delay operates:

```
1     module zdelay_tb();
2     // declare the counter signals
```

```
 3    reg clk;
 4    reg signed [7:0] zvalue;
 5    reg rst;
 6
 7    wire signed [7:0] gvalue;
 8    // Set up the initial variables and reset
 9    initial begin
10      $display ("time\t clk zvalue");
11      $monitor ("%g\t %b %d",
12      $time, clk, zvalue);
13      clk = 1;            // initialize the clock to 1
14      rst = 1;            // set the reset to 1 (not reset)
15      #5 rst = 0;         // reset = 0 : resets the counter
16      #10 rst = 1;        // reset back to 1 : counter can start
17      #10 zvalue = 23;
18      #10 zvalue = 54;
19      #10 zvalue = 12;
20      #1000 $finish;      // Finish the simulation
21    end
22
23    // Clock generator
24    always begin
25      $\mbox{\#}$5 clk = ~clk; // Clock every 5 time slots
26    end
27
28    zdelay $\mbox{\#}$(8) a1(clk, rst, zvalue, gvalue);
29
30    endmodule
```

The resulting behavior can clearly be seen in Figure 9.6.

Later in this book, the use of fixed and floating point numbers are discussed, as is the use of multiplication for more exact calculations and for practical filter design, where higher accuracy is required, it is likely that both these methods would be used. There may be situations, however, where it is simply not possible to use these advanced techniques,

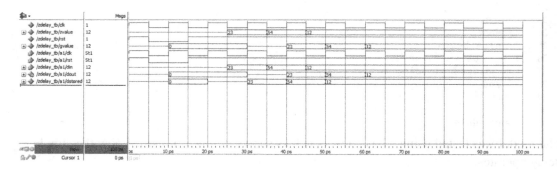

Figure 9.6
Z delay simulation.

particularly a problem when space is at a premium on the FPGA; in these cases, the simple approach described in this chapter will be required.

There are numerous texts on more advanced topics in digital filter design, and these are beyond the scope of this book, but it is useful to introduce some key concepts at this stage of the two main types of digital filter in common usage today. These are the recursive (or Infinite Impulse Response, IIR) filters and nonrecursive (or Finite Impulse Response, FIR) filters.

9.6 Finite Impulse Response Filters

Finite impulse response (FIR) filters are characterized by the fact that they use only delayed versions of the input signal to filter the input to the output. For example, if we take the expression for a general FIR filter below, we can see that the output is a function of a series of delayed, scaled versions of the input:

$$y = \sum_{i=0}^{n} A_i x[i] \tag{9.14}$$

where A_i is the scale factor for the ith delayed version of the input. We can represent this graphically in the diagram shown in Figure 9.7. We can implement this model using the basic building blocks described in this chapter of gain, division, sums and delays to develop block based models for such filters. As noted in the previous section, it is important to ensure that for higher accuracy filters, fixed or floating point arithmetic is required and also the use of multipliers for added accuracy is preferable in most cases to that of simple gain and division blocks as described previously in this chapter.

9.7 Infinite Impulse Response Filters

Infinite impulse response (IIR) filters are characterized by the fact that they use delayed versions of the input signal and fed-back and delayed versions of the output signal to filter the input to the output. For example, if we take the expression for a general IIR filter below, we can see that the output is a function of a series of delayed, scaled versions of the input and output.

$$y = \sum_{i=0}^{n} \frac{A_i x[i]}{B_i y[i]} \tag{9.15}$$

where A_i is the scale factor for the ith delayed version of the input and B_i is the scale factor for the ith delayed version of the output. This is obviously very similar to the FIR example previously given and can be built up using the same basic elements. If we consider the simple example earlier in this chapter, it can be seen that this is in fact a simple first order

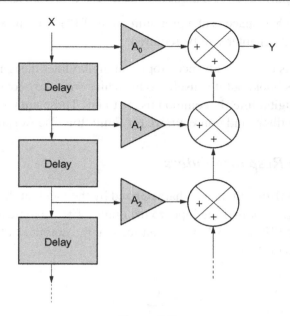

Figure 9.7
FIR filter schematic.

(single delay) IIR filter, with no delayed versions of the input and a single delayed version of
the output.

9.8 Summary

This chapter has introduced the concepts of implementing basic digital filters and Z-domain
functions in VHDL and Verilog and has given examples of both the building blocks and
constructed filters for implementation on an FPGA platform. The general concepts of FIR and
IIR filters have been introduced so that the reader can implement the topology and type of
filter appropriate for their own application. A detailed treatise on filters is beyond the scope of
this book and the reader is referred to one of the many digital filter design texts available for a
more in-depth analysis of the topic.

CHAPTER 10

Secure Systems

10.1 Introduction to Block Ciphers

The data encryption standard (DES) is a symmetric block cipher. A stream cipher operates on a digital data stream one or more bits at a time. A block cipher operates on complete blocks of data at any one time and produces a ciphertext block of equal size. DES is a block cipher that operates on data blocks of 64 bits in size. DES uses a 64-bit key 8×8 including 1 bit for parity, so the actual key is 56 bits. DES, in common with other block ciphers, is based around a structure called a *Feistel Lattice* so it is useful to describe how this works.

10.2 Feistel Lattice Structures

A block cipher operates on a plaintext block of n bits to produce a block of ciphertext of n bits. For the algorithm to be reversible (i.e., for decryption to be possible) there must be a unique mapping between the two sets of blocks. This can also be called a non singular transformation. For example, consider the following transformations as shown in Figure 10.1.

Obviously, this is essentially a substitution cipher, which may be susceptible to the standard statistical analysis techniques used for simple cryptanalysis of text (such as frequency analysis). As the block size increases, then this becomes increasingly less feasible. An obvious practical difficulty with this approach is the number of transformations required as n increases. This mapping is essentially the key and the number of bits will determine the key size. Therefore, for an n-bit general substitution block cipher, the key size is calculated as follows:

$$key = n \times 2^n \tag{10.1}$$

For a specific case where $n = 64$, the key size becomes $64 \times 2^{64} = 10^{21}$.

In order to get around this complexity problem, Feistel proposed an approach called a *product cipher* whereby the combination of several simple steps leads to a much more cryptographically secure solution than any of the component ciphers used. His approach relies on the alternation of two types of function:

- Diffusion
- Confusion

Design Recipes for FPGAs. http://dx.doi.org/10.1016/B978-0-08-097129-2.00010-6

	Reversible		Irreversible	
Plaintext	Ciphertext		Plaintext	Ciphertext
00	11		00	11
01	10		01	10
10	00		10	10
11	10		11	10

Figure 10.1
Reversible and irreversible transformations.

These two concepts are grounded in an approach developed by Shannon used in most standard block ciphers in common use today. Shannon's goal was to define cryptographic functions that would not be susceptible to statistical analysis. He therefore proposed two methods for reducing the ability of statistical cryptanalysis to find the original message, classified as diffusion and confusion.

In diffusion, the statistical structure of the plaintext is dissipated throughout the long-term statistics of the ciphertext. This is achieved by making each bit of the plaintext affect the value of many bits of the ciphertext. An example of this would be to add letters to a ciphertext such that the frequency of each letter is the same, regardless of the message. In binary block ciphers the technique uses multiple permutations and functions such that each bit of the ciphertext is affected by multiple bits in the plaintext.

Each block of plaintext is transformed into a block of ciphertext, and this depends on the key. Confusion aims to make the relationship between the ciphertext and the key as complex as possible to reduce the possibility of ascertaining the key. This requires a complex substitution algorithm, as a linear substitution would not protect the key.

Both diffusion and confusion are the cornerstones of successful block cipher design.

The result of these requirements is the Feistel Lattice (shown in Figure 10.2). This is the basic architecture that is used in block ciphers such as DES.

The inputs to the algorithm are the plaintext (of length 2w bits) and a key K. The plaintext is split into two halves L and R, and the data is then passed through n rounds of processing and then recombined to produce the ciphertext. Each round has an input L_{i-1} and R_{i-1} derived from the previous round and a subkey K_i, derived from the overall key K. Each round has the same structure. The left half of the data has a substitution performed. This requires a round function F to be performed on the right half of the data and then XORed with the left half. Finally a permutation is performed that requires the interchange of the two halves of the data.

The implementation of a Feistel network has the following key parameters:

- *Block size* A larger block size generally means greater security, but reduced speed. 64-bit block sizes are very heavily used as being a reasonable trade-off—although AES now uses 128 bits.

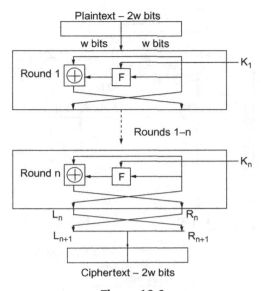

Figure 10.2
Feistel lattice structure.

- *Key Size* The same trade-off applies as for block size. Generally 64 bits is not now considered adequate and 128 bits is preferred.
- *Number of rounds* Each round adds additional security. A single round is inadequate, but 16 is considered standard.
- *Subkey generation* The more complex this algorithm is, the more secure the overall system will be.
- *Round function* Greater complexity again means greater resistance to cryptanalysis.

10.3 The Data Encryption Standard (DES)

10.3.1 Introduction

The Data Encryption Standard (DES) was adopted by the National Institute of Standards and Technology (NIST) in 1977 as the Federal Information Processing Standards 46 (FIPS PUB 46). As mentioned previously, the algorithm operates on plaintext blocks of 64 bits and the key size is 56 bits. By 1999, NIST had decreed that DES was no longer secure and should only be used for legacy systems and that triple DES should be used instead. As will be described later, DES has since been superceded by the Advanced Encryption Standards (AES). The coarse structure (overall architecture) of DES is shown in Figure 10.3.

The center section (where the main repetition occurs) is called the fine structure and is where the details of the encryption take place. This fine structure is detailed in Figure 10.4.

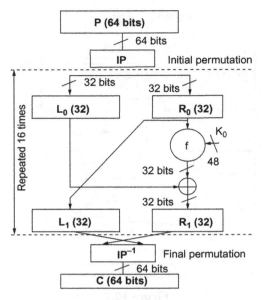

Figure 10.3
DES coarse structure.

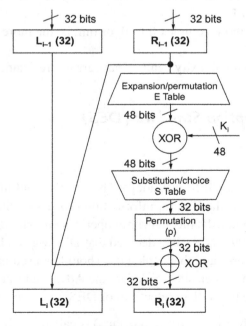

Figure 10.4
DES fine structure.

The fine structure of DES consists of several important functional blocks:

- *Initial permutation* Fixed, known mapping 64-64 bits.
- *Key transformations* Circular L shift of keys by A(i) bits in round (A(i) is known and fixed).
- *Compression Permutation* Fixed known subset of 56-bit input mapped onto 48-bit output.
- *Expansion permutation* 32-bit data shuffled and mapped (both operations fixed and known) onto 48 bits by duplicating 16 input bits. This makes diffusion quicker.

Another significant section of the algorithm is the substitution or S-box. The nonlinear aspect of the cipher is vital in cryptography. In DES the eight S boxes each contain four different (fixed and known) 4:4 input maps. These are selected by the extra bits created in the expansion box. The S boxes are structured as shown in Figure 10.5.

The final part of the DES structure is the key generation architecture for the individual round keys and this is given in Figure 10.6.

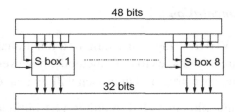

Figure 10.5
DES S-box architecture.

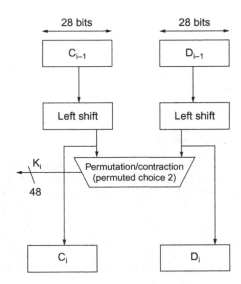

Figure 10.6
DES round key generation.

The remaining functional block is the initial and final permutation. The initial permutation (P-Box) is a 32:32 fixed, known bit permutation. The final permutation is the inverse of the initial permutation. The initial permutation is defined using the following table:

58	50	42	34	26	18	10	2
60	52	44	36	28	20	12	4
62	54	46	38	30	22	14	6
64	56	48	40	32	24	16	8
57	49	41	33	25	17	9	1
59	51	43	35	27	19	11	3
61	53	45	37	29	21	13	5
63	55	47	39	31	23	15	7

10.3.2 DES VHDL Implementation

DES can be implemented in VHDL using a structural or a functional approach. As has been discussed previously, there are advantages to both methods; however, the DES algorithm is tied implicitly to the structure, so a structural approach will give an efficient implementation.

Implementing the initial permutation in VHDL requires a 64-bit input vector and a 64-bit output vector. We can create this in VHDL with an entity that defines an input and output std_logic vector as follows. The architecture is simply the assignment of bits from input to output according to the initial permutation table previously defined.

```
1    library ieee;
2    use ieee.std_logic_1164.all;
3
4    entity des_ip is port
5    (
6     d : in  std_logic_vector(1 to 64);
7     y : out std_logic_vector(1 to 64)
8    );
9    end des_ip;
10
11   architecture behavior of des_ip is
12   begin
13     y(1)<=d(58);  y(2)<=d(50);  y(3)<=d(42);  y(4)<=d(34);
14     y(5)<=d(26);  y(6)<=d(18);  y(7)<=d(10);  y(8)<=d(2);
15     y(9)<=d(60);  y(10)<=d(52); y(11)<=d(44); y(12)<=d(36);
16     y(13)<=d(28); y(14)<=d(20); y(15)<=d(12); y(16)<=d(4);
17     y(17)<=d(62); y(18)<=d(54); y(19)<=d(46); y(20)<=d(38);
18     y(21)<=d(30); y(22)<=d(22); y(23)<=d(14); y(24)<=d(6);
19     y(25)<=d(64); y(26)<=d(56); y(27)<=d(48); y(28)<=d(40);
20     y(29)<=d(32); y(30)<=d(24); y(31)<=d(16); y(32)<=d(8);
21     y(33)<=d(57); y(34)<=d(49); y(35)<=d(41); y(36)<=d(33);
```

```
22      y(37)<=d(25); y(38)<=d(17); y(39)<=d(9);  y(40)<=d(1);
23       y(41)<=d(59); y(42)<=d(51); y(43)<=d(43); y(44)<=d(35);
24       y(45)<=d(27); y(46)<=d(19); y(47)<=d(11); y(48)<=d(3);
25       y(49)<=d(61); y(50)<=d(53); y(51)<=d(45); y(52)<=d(37);
26       y(53)<=d(29); y(54)<=d(21); y(55)<=d(13); y(56)<=d(5);
27       y(57)<=d(63); y(58)<=d(55); y(59)<=d(47); y(60)<=d(39);
28       y(61)<=d(31); y(62)<=d(23); y(63)<=d(15); y(64)<=d(7);
29      end behavior;
```

As this function is purely combinatorial we don't need to have a register (i.e., clocked input) on this model, although we could implement that if required using a simple process.

As shown in the previous description of the expansion function, we need to take a word consisting of 32 bits and expand it to 48 bits. This requires a translation table as shown below. Notice that there are duplicates in the cell which means that you only need 32 input bits to obtain 48 output bits.

32	1	2	3	4	5
4	5	6	7	8	9
8	9	10	11	12	13
12	13	14	15	16	17
16	17	18	19	20	21
20	21	22	23	24	25
24	25	26	27	28	29
28	29	30	31	32	1

We can use a VHDL model similar to the initial permutation function, except that in this case there are 32 input bits and 48 output bits. Notice that some of the input bits are repeated, giving a straightforward expansion function.

```
1      library ieee;
2      use ieee.std_logic_1164.all;
3
4      entity des_e is port
5      (
6        d : in  std_logic_vector(1 to 32);
7        y : out std_logic_vector(1 to 48)
8      );
9      end des_e;
```

The architecture is simply the assignment of bits from input to output according to the initial permutation table previously defined.

```
1      architecture behavior of des_e is
2      begin
3        y(1)<=d(32); y(2)<=d(1);  y(3)<=d(2);  y(4)<=d(3);
4        y(5)<=d(4);  y(6)<=d(5);  y(7)<=d(4);  y(8)<=d(5);
```

```
5      y(9)<=d(6);   y(10)<=d(7);   y(11)<=d(8);  y(12)<=d(9);
6      y(13)<=d(8);  y(14)<=d(9);   y(15)<=d(10); y(16)<=d(11);
7      y(17)<=d(12); y(18)<=d(13);  y(19)<=d(12); y(20)<=d(13);
8      y(21)<=d(14); y(22)<=d(15);  y(23)<=d(16); y(24)<=d(17);
9      y(25)<=d(16); y(26)<=d(17);  y(27)<=d(18); y(28)<=d(19);
10     y(29)<=d(20); y(30)<=d(21);  y(31)<=d(20); y(32)<=d(21);
11     y(33)<=d(22); y(34)<=d(23);  y(35)<=d(24); y(36)<=d(25);
12     y(37)<=d(24); y(38)<=d(25);  y(39)<=d(26); y(40)<=d(27);
13     y(41)<=d(28); y(42)<=d(29);  y(43)<=d(28); y(44)<=d(29);
14     y(45)<=d(30); y(46)<=d(31);  y(47)<=d(32); y(48)<=d(1);
15   end behavior;
```

The final permutation block is the permutation marked (P) on the fine structure after the key function. This is a straightforward bit substitution function with 32 bits input and 32 bits output. The bit translation table is shown in the following table:

16	7	20	21
29	12	28	17
1	15	23	26
5	18	31	10
2	8	24	14
32	27	3	9
19	13	30	6
22	11	4	25

This is implemented in VHDL using exactly the same approach as the previous expansion and permutation functions as follows:

```
1    library ieee;
2    use ieee.std_logic_1164.all;
3
4    entity des_p is port
5    (
6      d : in  std_logic_vector(1 to 32);
7      y : out std_logic_vector(1 to 32)
8    );
9    end des_p;
```

The architecture is simply the assignment of bits from input to output according to the initial permutation table previously defined.

```
1    architecture behavior of des_p is
2    begin
3      y(1)<=d(16);  y(2)<=d(7);   y(3)<=d(20);  y(4)<=d(21);
4      y(5)<=d(29);  y(6)<=d(12);  y(7)<=d(28);  y(8)<=d(17);
5      y(9)<=d(1);   y(10)<=d(15); y(11)<=d(23); y(12)<=d(26);
6      y(13)<=d(5);  y(14)<=d(18); y(15)<=d(31); y(16)<=d(10);
7      y(17)<=d(2);  y(18)<=d(8);  y(19)<=d(24); y(20)<=d(14);
```

```
8       y(21)<=d(32);  y(22)<=d(27);  y(23)<=d(3);   y(24)<=d(9);
9       y(25)<=d(19);  y(26)<=d(13);  y(27)<=d(30);  y(28)<=d(6);
10      y(29)<=d(22);  y(30)<=d(11);  y(31)<=d(4);   y(32)<=d(25);
11   end behavior;
```

The nonlinear part of the DES algorithm is the S box. This is a set of 6->4 bit transformations that reduce the 48 bits of the expanded word in the DES f function to the 32 bits for the next round. The required row and column are obtained from the data passed into the S box. The data into the S box is a 6-bit binary word. The row is obtained from $2.b1 + b6$ and the column is obtained from $b2b3b4b5$. For example, $S(011011)$ would give a row of 01 (1) and a column of 1101 (13). For S8 this would result in a value returning of 1110 (14).

The basic S-box entity can therefore be constructed using the following VHDL:

```
1    library ieee;
2    use ieee.std_logic_1164.all;
3    entity des_sbox is
4     port (
5      d : in  std_logic_vector (1 to 6);
6      y : out std_logic_vector (1 to 4)
7     );
8    end entity des_sbox;
```

One approach is to define the row and column from the input D word and then calculate the output Y word from that using a lookup table approach or minimize the logic as a truth table. The basic architecture could then look something like this:

```
1    architecture behavior of sbox is
2      signal r : std_logic_vector (1 to 2);
3      signal c : std_logic_vector (3 to 6);
4    begin
5      r <= d ( 1 to 2);
6      c <= d (3 to 6 );
7      -- the look up table or logic goes here
8    end;
```

Another approach is to define a simple lookup table with the input d as the unique address and the output y stored in the memory; this is exactly the same as a ROM, so the input is defined as an unsigned integer to look up the required value. In this case the memory is defined in exactly the same way as the ROM separately in this book.

The S-box substitutions are given in the table following and the VHDL can either use the lookup table approach to store the address of each substitution, or logic can be used to decode the correct output.

In order to use this table, the appropriate S box is selected and then the 2 bits of the row select the appropriate row and the same for the column. For example, for S box S1, if the row is 3

Row	Column Number															
	[0]	[1]	[2]	[3]	[4]	[5]	[6]	[7]	[8]	[9]	[10]	[11]	[12]	[13]	[14]	[15]
S1																
[0]	14	4	13	1	2	15	11	8	3	10	6	12	5	9	0	7
[1]	0	15	7	4	14	2	13	1	10	6	12	11	9	5	3	8
[2]	4	1	14	8	13	6	2	11	15	12	9	7	3	10	5	0
[3]	15	12	8	2	4	9	1	7	5	11	3	14	10	0	6	13
S2																
[0]	15	1	8	14	6	11	3	4	9	7	2	13	12	0	5	10
[1]	3	13	4	7	15	2	8	14	12	0	1	10	6	9	11	5
[2]	0	14	7	11	10	4	13	1	5	8	12	6	9	3	2	15
[3]	13	8	10	1	3	15	4	2	11	6	7	12	0	5	14	9
S3																
[0]	10	0	9	14	6	3	15	5	1	13	12	7	11	4	2	8
[1]	13	7	0	9	3	4	6	10	2	8	5	14	12	11	15	1
[2]	13	6	4	9	8	15	3	0	11	1	2	12	5	10	14	7
[3]	1	10	13	0	6	9	8	7	4	15	14	3	11	5	2	12
S4																
[0]	7	13	14	3	0	6	9	10	1	2	8	5	11	12	4	15
[1]	13	8	11	5	6	15	0	3	4	7	2	12	1	10	14	9
[2]	10	6	9	0	12	11	7	13	15	1	3	14	5	2	8	4
[3]	3	15	0	6	10	1	13	8	9	4	5	11	12	7	2	14
S5																
[0]	2	12	4	1	7	10	11	6	8	5	3	15	13	0	14	9
[1]	14	11	2	12	4	7	13	1	5	0	15	10	3	9	8	6
[2]	4	2	1	11	10	13	7	8	15	9	12	5	6	3	0	14
[3]	11	8	12	7	1	14	2	13	6	15	0	9	10	4	5	3
S6																
[0]	12	1	10	15	9	2	6	8	0	13	3	4	14	7	5	11
[1]	10	15	4	2	7	12	9	5	6	1	13	14	0	11	3	8
[2]	9	14	15	5	2	8	12	3	7	0	4	10	1	13	11	6
[3]	4	3	2	12	9	5	15	10	11	14	1	7	6	0	8	13
S7																
[0]	4	11	2	14	15	0	8	13	3	12	9	7	5	10	6	1
[1]	13	0	11	7	4	9	1	10	14	3	5	12	2	15	8	6
[2]	1	4	11	13	12	3	7	14	10	15	6	8	0	5	9	2
[3]	6	11	13	8	1	4	10	7	9	5	0	15	14	2	3	12
S8																
[0]	13	2	8	4	6	15	11	1	10	9	3	14	5	0	12	7
[1]	1	15	13	8	10	3	7	4	12	5	6	11	0	14	9	2
[2]	7	11	4	1	9	12	14	2	0	6	10	13	15	3	5	8
[3]	2	1	14	7	4	10	8	13	15	12	9	0	3	5	6	11

(11) and the Column is 10 (1010) then the output can be read off as 3 (0011). This can be coded in VHDL using nested case statements as follows:

```
1    case row is
2     when 0 =>
3       case column is
4         when 0 => y <= 14;
5         when 1 => y <= 4;
6         -- and so on
7       end case;
8     when 1 =>
9       case column is
10        -- and the lookup goes here
11      End case
12    -- and so on for all the rows
13    end case;
```

Obviously this is quite cumbersome, but also very easy to code automatically using a simple code generator and offers the possibility of the synthesis tool carrying out logic optimization and providing a much more efficient implementation than a memory block.

10.3.3 DES Verilog Implementation

DES can also be implemented in Verilog using a structural or a functional approach. As has been discussed previously, there are advantages to both methods; however, the DES algorithm is tied implicitly to the structure, so a structural approach will give an efficient implementation.

Implementing the initial permutation in Verilog requires a 64-bit input vector and a 64-bit output vector. In Verilog this is very easy to create by simply specifying the size of the input and output accordingly. In this combinatorial example (the same as the VHDL example previously in this chapter), we have not used an enable signal, but this would be easy to add, and then of course we can implement the output with a high impedance (z) state for unknown input values.

```
1    module desip (y,d);
2    // Define the IO
3    input [64:1] d;
4    output [64:1] y;
5
6    // Define the output as a register
7    reg [64:1] y;
8
9    // Assign the permutations on any change in d
10   always @ d
11   begin
```

```
12       y[1]<=d[58];    y[2]<=d[50];    y[3]<=d[42];    y[4]<=d[34];
13       y[5]<=d[26];    y[6]<=d[18];    y[7]<=d[10];    y[8]<=d[2];
14       y[9]<=d[60];    y[10]<=d[52];   y[11]<=d[44];   y[12]<=d[36];
15       y[13]<=d[28];   y[14]<=d[20];   y[15]<=d[12];   y[16]<=d[4];
16       y[17]<=d[62];   y[18]<=d[54];   y[19]<=d[46];   y[20]<=d[38];
17       y[21]<=d[30];   y[22]<=d[22];   y[23]<=d[14];   y[24]<=d[6];
18       y[25]<=d[64];   y[26]<=d[56];   y[27]<=d[48];   y[28]<=d[40];
19       y[29]<=d[32];   y[30]<=d[24];   y[31]<=d[16];   y[32]<=d[8];
20       y[33]<=d[57];   y[34]<=d[49];   y[35]<=d[41];   y[36]<=d[33];
21       y[37]<=d[25];   y[38]<=d[17];   y[39]<=d[9];    y[40]<=d[1];
22       y[41]<=d[59];   y[42]<=d[51];   y[43]<=d[43];   y[44]<=d[35];
23       y[45]<=d[27];   y[46]<=d[19];   y[47]<=d[11];   y[48]<=d[3];
24       y[49]<=d[61];   y[50]<=d[53];   y[51]<=d[45];   y[52]<=d[37];
25       y[53]<=d[29];   y[54]<=d[21];   y[55]<=d[13];   y[56]<=d[5];
26       y[57]<=d[63];   y[58]<=d[55];   y[59]<=d[47];   y[60]<=d[39];
27       y[61]<=d[31];   y[62]<=d[23];   y[63]<=d[15];   y[64]<=d[7];
28       end
29       endmodule
```

As this function is purely combinatorial we don't need to have a register (i.e., clocked input) on this model, although we could implement that if required using a simple addition of a clock.

As shown in the previous description of the expansion function, we need to take a word consisting of 32 bits and expand it to 48 bits. This requires a translation table as shown below. Notice that there are duplicates in the cell, which means that you only need 32 input bits to obtain 48 output bits.

32	1	2	3	4	5
4	5	6	7	8	9
8	9	10	11	12	13
12	13	14	15	16	17
16	17	18	19	20	21
20	21	22	23	24	25
24	25	26	27	28	29
28	29	30	31	32	1

We can use a Verilog model similar to the initial permutation function, except that in this case there are 32 input bits and 48 output bits. Notice that some of the input bits are repeated, giving a straightforward expansion function.

```
1        module dese (y,d);
2        // Define the IO
3        input [32:1] d;
4        output [48:1] y;
5
6        // Define the output as a register
```

```
7    reg [48:1] y;
8
9    // Assign the permutations on any change in d
10   always @ d
11   begin
12     y[1]<=d[32];  y[2]<=d[1];   y[3]<=d[2];   y[4]<=d[3];
13     y[5]<=d[4];   y[6]<=d[5];   y[7]<=d[4];   y[8]<=d[5];
14     y[9]<=d[6];   y[10]<=d[7];  y[11]<=d[8];  y[12]<=d[9];
15     y[13]<=d[8];  y[14]<=d[9];  y[15]<=d[10]; y[16]<=d[11];
16     y[17]<=d[12]; y[18]<=d[13]; y[19]<=d[12]; y[20]<=d[13];
17     y[21]<=d[14]; y[22]<=d[15]; y[23]<=d[16]; y[24]<=d[17];
18     y[25]<=d[16]; y[26]<=d[17]; y[27]<=d[18]; y[28]<=d[19];
19     y[29]<=d[20]; y[30]<=d[21]; y[31]<=d[20]; y[32]<=d[21];
20     y[33]<=d[22]; y[34]<=d[23]; y[35]<=d[24]; y[36]<=d[25];
21     y[37]<=d[24]; y[38]<=d[25]; y[39]<=d[26]; y[40]<=d[27];
22     y[41]<=d[28]; y[42]<=d[29]; y[43]<=d[28]; y[44]<=d[29];
23     y[45]<=d[30]; y[46]<=d[31]; y[47]<=d[32]; y[48]<=d[1];
24   end
25   endmodule
```

The final permutation block is the permutation marked (P) on the fine structure after the key function. This is a straightforward bit substitution function with 32 bits input and 32 bits output. The bit translation is shown in the following table:

16	7	20	21
29	12	28	17
1	15	23	26
5	18	31	10
2	8	24	14
32	27	3	9
19	13	30	6
22	11	4	25

This is implemented in Verilog using exactly the same approach as the previous expansion and permutation functions as follows, with the behavior simply the assignment of bits from input to output according to the initial permutation table previously defined.

```
1    module desp (y,d);
2    // Define the IO
3    input [32:1] d;
4    output [32:1] y;
5
6    // Define the output as a register
7    reg [32:1] y;
8
9    // Assign the permutations on any change in d
10   always @ d
11   begin
```

```
12    y[1]<=d[16];   y[2]<=d[7];   y[3]<=d[20];   y[4]<=d[21];
13    y[5]<=d[29];   y[6]<=d[12];  y[7]<=d[28];   y[8]<=d[17];
14    y[9]<=d[1];    y[10]<=d[15]; y[11]<=d[23];  y[12]<=d[26];
15    y[13]<=d[5];   y[14]<=d[18]; y[15]<=d[31];  y[16]<=d[10];
16    y[17]<=d[2];   y[18]<=d[8];  y[19]<=d[24];  y[20]<=d[14];
17    y[21]<=d[32];  y[22]<=d[27]; y[23]<=d[3];   y[24]<=d[9];
18    y[25]<=d[19];  y[26]<=d[13]; y[27]<=d[30];  y[28]<=d[6];
19    y[29]<=d[22];  y[30]<=d[11]; y[31]<=d[4];   y[32]<=d[25];
20  end
21  endmodule
```

The nonlinear part of the DES algorithm is the S box. This is a set of 6->4 bit transformations that reduce the 48 bits of the expanded word in the DES f function to the 32 bits for the next round. The required row and column are obtained from the data passed into the S box. The data into the S box is a 6-bit binary word. The row is obtained from $2.b1 + b6$ and the column is obtained from $b2b3b4b5$. For example, S(011011) would give a row of 01 (1) and a column of 1101 (13). For S8 this would result in a value returning of 1110 (14).

The basic S-box model can therefore be constructed using the following Verilog, with one approach to define the row and column from the input d word and then calculate the output Y word from that using a lookup table approach or minimize the logic as a truth table. The basic architecture could then look something like this:

```
1   module sbox (y,d);
2   // Define the IO
3   input [6:1] d;
4   output [4:1] y;
5
6   // Define the output as a register
7   reg [4:1] y;
8
9   // Assign the permutations on any change in d
10  reg [2:1] r;
11  reg [4:1] c;
12
13  always @ d
14  begin
15    r <= d[2:1];
16    c <= d[6:3];
17
18    // The remainder of the SBOX logic goes here
19
20  end
21  endmodule
```

Another approach is to define a simple lookup table with the input d as the unique address and the output y stored in the memory; this is exactly the same as a ROM, so the input is defined as an unsigned integer to look up the required value. In this case the memory is defined in exactly the same way as the ROM is defined separately in this book.

In order to use the table, the appropriate S box is selected and then the 2 bits of the row select the appropriate row and the same for the column. For example, for Sbox S1, if the row is 3 (11) and the Column is 10 (1010) then the output can be read off as 3 (0011). This can be coded in VHDL using nested case statements as follows:

```
1    case (row)
2      0: case (column)
3        0: y=14;
4        1: y=4;
5        // and so on
6      endcase
7      1: case (column)
8        0: y=0;
9        1: y=15;
10       // and so on
11     endcase
12     // and so on for all the rows
13   endcase
```

Obviously this is quite cumbersome, but also very easy to code automatically using a simple code generator and offers the possibility of the synthesis tool carrying out logic optimization and providing a much more efficient implementation than a memory block.

10.3.4 Validation of DES

In order to validate the implementation of DES, a set of test vectors can be used (i.e., plaintext/ciphertext pairs to ensure that the correct processing is taking place). A suitable set of test vectors is given as:

Plaintext	Ciphertext
4E6F772069732074	3FA40E8A984D4815
68652074696D6520	6A271787AB8883F9
666F7220616C6C20	893D51EC4B563B53

In this case the key to be used is 0123456789ABCDEF.

Each of the groups of hexadecimal characters is represented by 7-bit ASCII and adding an extra bit.

10.4 Advanced Encryption Standard

In 1997, the U.S. National Institute of Standards and Technology (NIST) published a request for information regarding the creation of a new Advanced Encryption Standard (AES) for nonclassified government documents. The call also stipulated that the AES would specify an unclassified, publicly disclosed encryption algorithm(s), available royalty-free, worldwide. In addition, the algorithm(s) must implement symmetric key cryptography as a block cipher and (at a minimum) support block sizes of 128 bits and key sizes of 128, 192, and 256 bits.

After an open competition, the Rijndael algorithm was chosen as the winner and implemented as the AES standard. Rijndael allows key and block sizes to be 128, 192, or 256 bits. AES allows the same key sizes, but operates using a block size of 128 bits. The algorithm operates in a similar way to DES, with 10 rounds of confusion and diffusion operators (shuffling and mixing) blocks at a time. Each round has a separate key, generated from the overall key. The round structure is shown in Figure 10.7.

The overall AES structure is given in Figure 10.8.

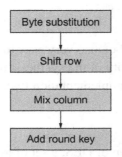

Figure 10.7
AES round structure.

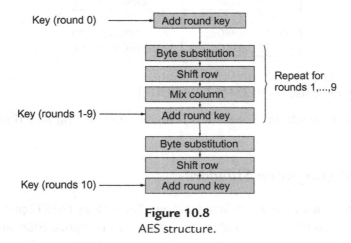

Figure 10.8
AES structure.

Each block consists of 128 bits, and these are divided into 16 8-bit bytes. Each of the operations acts upon these 8-bit bytes in a 4×4 matrix:

$$\begin{pmatrix} a_{0,0} & a_{0,1} & a_{0,2} & a_{0,3} \\ a_{1,0} & a_{1,1} & a_{1,2} & a_{1,3} \\ a_{2,0} & a_{2,1} & a_{2,2} & a_{2,3} \\ a_{3,0} & a_{3,1} & a_{3,2} & a_{3,3} \end{pmatrix}$$

Note that each element $(a_{i,j})$ is an 8-bit byte, viewed as elements of $GF(2^8)$. The arithmetic operators take advantage of the Galois Field (GF) rules defined in the Rijndael algorithm; an example is that of addition that is implemented by XOR.

Multiplication is more complicated, but each byte has the multiplicative inverse such that $b.b' = 00000001$ (apart from 00000000, whose multiplicative inverse is 00000000). The model of the finite field $GF(2^8)$ depends on the choice of an irreducible polynomial of degree 8, which for Rijndael is:

$$X^8 + X^4 + X^3 + 1 \tag{10.2}$$

Each of the round operations requires a specific mathematical exploration. Taking each in turn we can establish the requirements for each one.

Taking the original matrix:

$$\begin{pmatrix} a_{0,0} & a_{0,1} & a_{0,2} & a_{0,3} \\ a_{1,0} & a_{1,1} & a_{1,2} & a_{1,3} \\ a_{2,0} & a_{2,1} & a_{2,2} & a_{2,3} \\ a_{3,0} & a_{3,1} & a_{3,2} & a_{3,3} \end{pmatrix}$$

each element can be replaced byte-by-byte to generate a new matrix:

$$\begin{pmatrix} b_{0,0} & b_{0,1} & b_{0,2} & b_{0,3} \\ b_{1,0} & b_{1,1} & b_{1,2} & b_{1,3} \\ b_{2,0} & b_{2,1} & b_{2,2} & b_{2,3} \\ b_{3,0} & b_{3,1} & b_{3,2} & b_{3,3} \end{pmatrix}$$

Byte substitution requires that for each input data block $a(3, 3)$, we look up a table of substitutions and replace the bytes to produce a new matrix $b(3, 3)$. The way it works is that for each input byte abcdefgh, we look up row abcd and column efgh and use the byte at that location in the output.

The complete byte substitution table is defined as shown in Figure 10.9.

For example: If the input data byte was 7A, then this in binary terms is:

0111 1010

So the row required is 7 (0111) and the column required is A (1010). From this we can read off the resulting number from the table:

$218 = 1101\ 1010 = DA$

This is illustrated in the byte substitution table in Figure 10.9.

We can see that this is a bit shuffling operation that is simply moving bytes around in a publicly defined manner that does not have anything to do with a key.

Also note that the individual bits within the byte are not changed as such. This is a bytewise operation. The Shift Row function is essentially a set of cyclic shifts to the left with offsets of 0, 1, 2, 3, respectively.

$$
\begin{pmatrix}
c_{0,0} & c_{0,1} & c_{0,2} & c_{0,3} \\
c_{1,0} & c_{1,1} & c_{1,2} & c_{1,3} \\
c_{2,0} & c_{2,1} & c_{2,2} & c_{2,3} \\
c_{3,0} & c_{3,1} & c_{3,2} & c_{3,3}
\end{pmatrix}
=
\begin{pmatrix}
b_{0,0} & b_{0,1} & b_{0,2} & b_{0,3} \\
b_{1,1} & b_{1,2} & b_{1,3} & b_{1,0} \\
b_{2,2} & b_{2,3} & b_{2,0} & b_{2,1} \\
b_{3,3} & b_{3,0} & b_{3,1} & b_{3,2}
\end{pmatrix}
$$

	0	1	2	3	4	5	6	7	8	9	A	B	C	D	E	F
0	099	124	119	123	242	107	111	197	048	001	103	043	254	215	171	118
1	202	130	201	125	250	089	071	240	173	212	162	175	156	164	114	192
2	183	253	147	038	054	063	247	204	052	165	229	241	113	216	049	021
3	004	199	035	195	024	150	005	154	007	018	128	226	235	039	178	117
4	009	131	044	026	027	110	090	160	082	059	214	179	041	227	047	132
5	083	209	000	237	032	252	177	091	106	203	190	057	074	076	088	207
6	208	239	170	251	067	077	051	133	069	249	002	127	080	060	159	168
7	081	163	064	143	146	157	056	245	188	182	218	033	016	255	243	210
8	205	012	019	236	095	151	068	023	196	167	126	061	100	093	025	115
9	096	129	079	220	034	042	144	136	070	238	184	020	222	094	011	219
A	224	050	058	010	073	006	036	092	194	211	172	098	145	149	228	121
B	231	200	055	109	141	213	078	169	108	086	244	234	101	122	174	008
C	186	120	037	046	028	166	180	198	232	221	116	031	075	189	139	138
D	112	062	181	102	072	003	246	014	097	053	087	185	134	193	029	158
E	225	248	152	017	105	217	142	148	155	030	135	233	206	085	040	223
F	140	161	137	013	191	230	066	104	065	153	045	015	176	084	187	022

Figure 10.9
AES byte substitution table.

The Mix Columns function is a series of specific multiplications:

$$
\begin{pmatrix}
d_{0,0} & d_{0,1} & d_{0,2} & d_{0,3} \\
d_{1,0} & d_{1,1} & d_{1,2} & d_{1,3} \\
d_{2,0} & d_{2,1} & d_{2,2} & d_{2,3} \\
d_{3,0} & d_{3,1} & d_{3,2} & d_{3,3}
\end{pmatrix}
=
\begin{pmatrix}
02 & 03 & 01 & 01 \\
01 & 02 & 03 & 01 \\
01 & 01 & 02 & 03 \\
03 & 01 & 01 & 02
\end{pmatrix}
*
\begin{pmatrix}
c_{0,0} & c_{0,1} & c_{0,2} & c_{0,3} \\
c_{1,0} & c_{1,1} & c_{1,2} & c_{1,3} \\
c_{2,0} & c_{2,1} & c_{2,2} & c_{2,3} \\
c_{3,0} & c_{3,1} & c_{3,2} & c_{3,3}
\end{pmatrix}
$$

where $01 = 00000001$, $02 = 00000010$, $03 = 00000011$.

All multiplications are $GF(2^8)$ and this transformation is invertible.

The final operation in each round is to add the key using the following function:

$$
\begin{pmatrix}
e_{0,0} & e_{0,1} & e_{0,2} & e_{0,3} \\
e_{1,0} & e_{1,1} & e_{1,2} & e_{1,3} \\
e_{2,0} & e_{2,1} & e_{2,2} & e_{2,3} \\
e_{3,0} & e_{3,1} & e_{3,2} & e_{3,3}
\end{pmatrix}
=
\begin{pmatrix}
d_{0,0} & d_{0,1} & d_{0,2} & d_{0,3} \\
d_{1,0} & d_{1,1} & d_{1,2} & d_{1,3} \\
d_{2,0} & d_{2,1} & d_{2,2} & d_{2,3} \\
d_{3,0} & d_{3,1} & d_{3,2} & d_{3,3}
\end{pmatrix}
\oplus
\begin{pmatrix}
k_{0,0} & k_{0,1} & k_{0,2} & k_{0,3} \\
k_{1,0} & k_{1,1} & k_{1,2} & k_{1,3} \\
k_{2,0} & k_{2,1} & k_{2,2} & k_{2,3} \\
k_{3,0} & k_{3,1} & k_{3,2} & k_{3,3}
\end{pmatrix}
$$

The round keys are generated using the following method. The original key of 128 bits is represented as a 4×4 matrix of bytes (of 8 bits). This can be thought of as four columns $W(0), W(1), W(2), W(3)$. Adjoin 40 columns $W(4), \ldots, W(43)$. The round key for round i consists of i columns. If i is a multiple of 4:

$$W(i) = W(i - 4) \oplus T(W(i - 1)) \tag{10.3}$$

where T is a transformation of a, b, c, d in column $W(i - 1)$ using the following procedure:

- Shift cyclically to get b, c, d, a.
- Replace each byte with S-box entry using ByteSub, to get e, f, g, h.
- Compute round constant $r(i) = 00000010(i - 4)/4$ in $GF(2^8)$.
- $T(W(i - 1)) = (e \oplus r(i), f, g, h)$
- If i is not a multiple of 4, $W(i) = W(i - 4) \oplus W(i - 1)$

10.4.1 Implementing AES in VHDL

We have two options for implementing block cipher operations in VHDL. We can use the structural approach (shown in the DES example previously in this chapter), or sometimes it makes sense to define a library of functions and use those to make much simpler models.

In the example of AES, we can define a top level entity and architecture that has the bare minimum of structure and is completely defined using functions. This can be especially useful when working with behavioral synthesis software as this allows complete flexibility for architectural optimization.

```
1    library ieee;
2    use ieee.std_logic_1164.all;
3    entity AES is
4      port(
5        plaintext : in std_logic_vector(127 downto 0);
6        keytext : in std_logic_vector(127 downto 0);
7        encrypt : in std_logic;
8        go : in std_logic;
9        ciphertext : out std_logic_vector(127 downto 0);
10       done : out std_logic := '0'
11     );
12   end;
13
14   use work.aes_functions.all;
15   architecture behavior of AES is
16   begin
17     process
18     begin
19       wait until go = '1';
20       done <= '0';
21       ciphertext <= aes_core(plaintext, keytext, encrypt);
22       done <= '1';
23     end process;
24   end;
```

In this example, we have the plaintext and keytext inputs defined as 128-bit-wide vectors and the ciphertext output is also defined as 128 bits wide. The go flag initiates the encryption and the done flag shows when this has been completed.

Notice that we have a work library defined called aes_functions which encapsulates all the relevant functions for the AES algorithm. The set of functions is defined in a package (aes_functions) and the VHDL is given:

```
1    library ieee;
2    use ieee.std_logic_1164.all;
3    use ieee.numeric_std.all;
4    package aes_functions is
5
6      constant nr : integer := 10;
7      constant nb : integer := 4;
8      constant nk : integer := 4;
9
10     subtype vec1408 is std_logic_vector(1407 downto 0);
11     subtype vec128 is std_logic_vector(127 downto 0);
12     subtype vec64 is std_logic_vector(63 downto 0);
13     subtype vec32 is std_logic_vector(31 downto 0);
14     subtype vec16 is std_logic_vector(15 downto 0);
15     subtype vec8 is std_logic_vector(7 downto 0);
16
17     subtype int9 is integer range 0 to 9;
18
19     function input_output (input : vec128) return vec128;
20     function sbox (pt : vec8) return vec8;
```

```
21    function subbytes (plaintext : vec128) return vec128;
22    function shiftrows (plaintext : vec128) return vec128;
23    function ffmul(pt : vec8; mul : vec8) return vec8;
24    function mixcl( 10 : vec8; 11 : vec8; 12 : vec8; 13 : vec8) return vec8;
25    function mixcolumns(pt : vec128) return vec128;
26    function rcon (input : int9) return vec8;
27    function aes_keyexpansion(key : vec128) return vec1408;
28    function aes_core (signal plaintext : vec128; signal keytext : vec128;
         signal encrypt : std_logic) return vec128;
29
30
31
32    end;
33
34    library ieee;
35    use ieee.std_logic_1164.all;
36    use ieee.numeric_std.all;
37    package body aes_functions is
38
39    ------------------------------------------------------------------------
40    function subbytes (plaintext : vec128)
41    -- moods inline
42    return vec128 is
43      variable ciphertext : vec128;
44    begin
45      ciphertext := sbox(plaintext(127 downto 120)) &
46                    sbox(plaintext(119 downto 112)) &
47                    sbox(plaintext(111 downto 104)) &
48                    sbox(plaintext(103 downto 96)) &
49                    sbox(plaintext(95  downto 88)) &
50                    sbox(plaintext(87  downto 80)) &
51                    sbox(plaintext(79  downto 72)) &
52                    sbox(plaintext(71  downto 64)) &
53                    sbox(plaintext(63  downto 56)) &
54                    sbox(plaintext(55  downto 48)) &
55                    sbox(plaintext(47  downto 40)) &
56                    sbox(plaintext(39  downto 32)) &
57                    sbox(plaintext(31  downto 24)) &
58                    sbox(plaintext(23  downto 16)) &
59                    sbox(plaintext(15  downto 8)) &
60                    sbox(plaintext(7   downto 0));
61      return ciphertext;
62    end;
63
64    ------------------------------------------------------------------------
65    function shiftrows (plaintext : vec128)
66    -- moods inline
67    return vec128 is
68      variable ciphertext : vec128;
69    begin
70      --line 0 (the first): no shift
71      ciphertext := plaintext(31  downto 24) &
72                    plaintext(55  downto 48) &
73                    plaintext(79  downto 72) &
74                    plaintext(103 downto 96) &
```

```
75                      plaintext(127 downto 120) &
76                      plaintext(23  downto 16) &
77                      plaintext(47  downto 40) &
78                      plaintext(71  downto 64) &
79                      plaintext(95  downto 88) &
80                      plaintext(119 downto 112) &
81                      plaintext(15  downto 8) &
82                      plaintext(39  downto 32) &
83                      plaintext(63  downto 56) &
84                      plaintext(87  downto 80) &
85                      plaintext(111 downto 104) &
86                      plaintext(7   downto 0);
87        return ciphertext;
88      end;
89
90    ---------------------------------------------------------------
91
92    ---------------------------------------------------------------
93      function tablelog (input : vec8)
94        -- moods inline
95      return vec8 is
96        variable output : vec8;
97        type table256 is array(0 to 255) of vec8;
98        constant pt_256 : table256 := (
99        -- moods rom
100          x"00", x"00", x"19", x"01", x"32",
101          x"02", x"1a", x"c6", x"4b",
102          x"c7", x"1b", x"68", x"33",
103          x"ee", x"df", x"03", x"64",
104
105          x"04", x"e0", x"0e", x"34",
106          x"8d", x"81", x"ef", x"4c",
107          x"71", x"08", x"c8", x"f8",
108          x"69", x"1c", x"c1", x"7d",
109
110          x"c2", x"1d", x"b5", x"f9",
111          x"b9", x"27", x"6a", x"4d",
112          x"e4", x"a6", x"72", x"9a",
113          x"c9", x"09", x"78", x"65",
114
115          x"2f", x"8a", x"05", x"21",
116          x"0f", x"e1", x"24", x"12",
117          x"f0", x"82", x"45", x"35",
118          x"93", x"da", x"8e", x"96",
119
120          x"8f", x"db", x"bd", x"36",
121          x"d0", x"ce", x"94", x"13",
122          x"5c", x"d2", x"f1", x"40",
123          x"46", x"83", x"38", x"66",
124          x"dd", x"fd", x"30", x"bf",
125          x"06", x"8b", x"62", x"b3",
126          x"25", x"e2", x"98", x"22",
127          x"88", x"91", x"10", x"7e",
128
129          x"6e", x"48", x"c3", x"a3",
```

```
130          x"b6", x"1e", x"42", x"3a",
131          x"6b", x"28", x"54", x"fa",
132          x"85", x"3d", x"ba", x"2b",
133
134          x"79", x"0a", x"15", x"9b",
135          x"9f", x"5e", x"ca", x"4e",
136          x"d4", x"ac", x"e5", x"f3",
137          x"73", x"a7", x"57", x"af",
138
139          x"58", x"a8", x"50", x"f4",
140          x"ea", x"d6", x"74", x"4f",
141          x"ae", x"e9", x"d5", x"e7",
142          x"e6", x"ad", x"e8", x"2c",
143
144          x"d7", x"75", x"7a", x"eb",
145          x"16", x"0b", x"f5", x"59",
146          x"cb", x"5f", x"b0", x"9c",
147          x"a9", x"51", x"a0", x"7f",
148
149          x"0c", x"f6", x"6f", x"17",
150          x"c4", x"49", x"ec", x"d8",
151          x"43", x"1f", x"2d", x"a4",
152          x"76", x"7b", x"b7", x"cc",
153
154          x"bb", x"3e", x"5a", x"fb",
155          x"60", x"b1", x"86", x"3b",
156          x"52", x"a1", x"6c", x"aa",
157          x"55", x"29", x"9d", x"97",
158
159          x"b2", x"87", x"90", x"61",
160          x"be", x"dc", x"fc", x"bc",
161          x"95", x"cf", x"cd", x"37",
162          x"3f", x"5b", x"d1", x"53",
163
164          x"39", x"84", x"3c", x"41",
165          x"a2", x"6d", x"47", x"14",
166          x"2a", x"9e", x"5d", x"56",
167          x"f2", x"d3", x"ab", x"44",
168
169          x"11", x"92", x"d9", x"23",
170          x"20", x"2e", x"89", x"b4",
171          x"7c", x"b8", x"26", x"77",
172          x"99", x"e3", x"a5", x"67",
173
174          x"4a", x"ed", x"de", x"c5",
175          x"31", x"fe", x"18", x"0d",
176          x"63", x"8c", x"80", x"c0",
177          x"f7", x"70", x"07" );
178    begin
179      output := pt_256(to_integer(unsigned(input)));
180      return output;
181    end;
182
183    ----------------------------------------------------------------
184    function tableexp (input : vec8)
```

```
185          -- moods inline
186      return vec8 is
187       variable output : vec8;
188       type table256 is array(0 to 255) of vec8;
189       constant pt_256 : table256 := (
190       -- moods rom
191        x"01", x"03", x"05", x"0f",
192        x"11", x"33", x"55", x"ff",
193        x"1a", x"2e", x"72", x"96",
194        x"a1", x"f8", x"13", x"35",
195
196        x"5f", x"e1", x"38", x"48",
197        x"d8", x"73", x"95", x"a4",
198        x"f7", x"02", x"06", x"0a",
199        x"1e", x"22", x"66", x"aa",
200
201        x"e5", x"34", x"5c", x"e4",
202        x"37", x"59", x"eb", x"26",
203        x"6a", x"be", x"d9", x"70",
204        x"90", x"ab", x"e6", x"31",
205
206        x"53", x"f5", x"04", x"0c",
207        x"14", x"3c", x"44", x"cc",
208        x"4f", x"d1", x"68", x"b8",
209        x"d3", x"6e", x"b2", x"cd",
210
211        x"4c", x"d4", x"67", x"a9",
212        x"e0", x"3b", x"4d", x"d7",
213        x"62", x"a6", x"f1", x"08",
214        x"18", x"28", x"78", x"88",
215
216        x"83", x"9e", x"b9", x"d0",
217        x"6b", x"bd", x"dc", x"7f",
218        x"81", x"98", x"b3", x"ce",
219        x"49", x"db", x"76", x"9a",
220
221        x"b5", x"c4", x"57", x"f9",
222        x"10", x"30", x"50", x"f0",
223        x"0b", x"1d", x"27", x"69",
224        x"bb", x"d6", x"61", x"a3",
225
226        x"fe", x"19", x"2b", x"7d",
227        x"87", x"92", x"ad", x"ec",
228        x"2f", x"71", x"93", x"ae",
229        x"e9", x"20", x"60", x"a0",
230
231        x"fb", x"16", x"3a", x"4e",
232        x"d2", x"6d", x"b7", x"c2",
233        x"5d", x"e7", x"32", x"56",
234        x"fa", x"15", x"3f", x"41",
235
236        x"c3", x"5e", x"e2", x"3d",
237        x"47", x"c9", x"40", x"c0",
238        x"5b", x"ed", x"2c", x"74",
239        x"9c", x"bf", x"da", x"75",
```

```
240
241              x"9f", x"ba", x"d5", x"64",
242              x"ac", x"ef", x"2a", x"7e",
243              x"82", x"9d", x"bc", x"df",
244              x"7a", x"8e", x"89", x"80",
245
246              x"9b", x"b6", x"c1", x"58",
247              x"e8", x"23", x"65", x"af",
248              x"ea", x"25", x"6f", x"b1",
249              x"c8", x"43", x"c5", x"54",
250
251              x"fc", x"1f", x"21", x"63",
252              x"a5", x"f4", x"07", x"09",
253              x"1b", x"2d", x"77", x"99",
254              x"b0", x"cb", x"46", x"ca",
255
256              x"45", x"cf", x"4a", x"de",
257              x"79", x"8b", x"86", x"91",
258              x"a8", x"e3", x"3e", x"42",
259              x"c6", x"51", x"f3", x"0e",
260
261              x"12", x"36", x"5a", x"ee",
262              x"29", x"7b", x"8d", x"8c",
263              x"8f", x"8a", x"85", x"94",
264              x"a7", x"f2", x"0d", x"17",
265
266              x"39", x"4b", x"dd", x"7c",
267              x"84", x"97", x"a2", x"fd",
268              x"1c", x"24", x"6c", x"b4",
269              x"c7", x"52", x"f6", x"01");
270        begin
271          output := pt_256(to_integer(unsigned(input)));
272        return output;
273        end;
274
275      --------------------------------------------------------------------
276        function ffmul(pt : vec8; mul : vec8)
277        -- moods inline
278        return vec8 is
279          --variable res : vec8;
280          variable tablogpt : vec8;
281          variable tablogmul : vec8;
282          variable tablogpt8 : unsigned(8 downto 0);
283          variable tablogmul8 : unsigned(8 downto 0);
284          variable carrie : std_logic_vector(8 downto 0);
285          variable power : vec8;
286          variable result: vec8;
287        begin
288          tablogpt := tablelog(pt);
289          tablogmul := tablelog(mul);
290
291          tablogpt8 := unsigned("0" & tablogpt);
292          tablogmul8 := unsigned("0" & tablogmul);
293
294          carrie := std_logic_vector(tablogmul8 + tablogpt8);
```

```
295
296          if pt = x"00" or mul = x"00" then
297            result := x"00";
298          elsif carrie(8) = '1' or carrie(7 downto 0) = x"ff" then -- mod 255
299            power := std_logic_vector(unsigned(carrie(7 downto 0)) + 1); -- power =
                   power - 255
300            result := tableexp(power);
301          else
302            power := carrie(7 downto 0);
303            result := tableexp(power);
304          end if;
305          return result;
306        end;
307
308    -----------------------------------------------------------------
309      function mixcl( l0 : vec8; l1 : vec8; l2 : vec8; l3 : vec8)
310        -- moods inline
311      return vec8 is
312        variable ct : vec8;
313      begin
314
315        ct := ffmul(l0, x"02") xor ffmul(l1, x"01") xor ffmul(l2, x"01") xor ffmul
               (l3, x"03");
316
317        return ct;
318      end;
319
320    -----------------------------------------------------------------
321      function mixcolumns(pt : vec128)
322        -- moods inline
323      return vec128 is
324        variable ct : vec128;
325      begin
326
327        ct := mixcl(pt(127 downto 120), pt(119 downto 112), pt(111 downto 104), pt
               (103 downto 96)) &
328            mixcl(pt(119 downto 112), pt(111 downto 104), pt(103 downto 96), pt
                   (127 downto 120)) &
329            mixcl(pt(111 downto 104), pt(103 downto 96), pt(127 downto 120), pt
                   (119 downto 112)) &
330            mixcl(pt(103 downto 96), pt(127 downto 120), pt(119 downto 112), pt
                   (111 downto 104)) &
331
332            mixcl(pt(95 downto 88), pt(87 downto 80), pt(79 downto 72), pt(71
                   downto 64)) &
333            mixcl(pt(87 downto 80), pt(79 downto 72), pt(71 downto 64), pt(95
                   downto 88)) &
334            mixcl(pt(79 downto 72), pt(71 downto 64), pt(95 downto 88), pt(87
                   downto 80)) &
335            mixcl(pt(71 downto 64), pt(95 downto 88), pt(87 downto 80), pt(79
                   downto 72)) &
336
337            mixcl(pt(63 downto 56), pt(55 downto 48), pt(47 downto 40), pt(39
                   downto 32)) &
```

```
338         mixcl(pt(55 downto 48), pt(47 downto 40), pt(39 downto 32), pt(63
                  downto 56)) &
339         mixcl(pt(47 downto 40), pt(39 downto 32), pt(63 downto 56), pt(55
                  downto 48)) &
340         mixcl(pt(39 downto 32), pt(63 downto 56), pt(55 downto 48), pt(47
                  downto 40)) &
341
342         mixcl(pt(31 downto 24), pt(23 downto 16), pt(15 downto 8), pt(7 downto
                  0)) &
343         mixcl(pt(23 downto 16), pt(15 downto 8), pt(7 downto 0), pt(31 downto
                  24)) &
344         mixcl(pt(15 downto 8), pt(7 downto 0), pt(31 downto 24), pt(23 downto
                  16)) &
345         mixcl(pt(7 downto 0), pt(31 downto 24), pt(23 downto 16), pt(15 downto
                  8));
346
347     return ct;
348   end;
349
350   -----------------------------------------------------------------
351   function input_output (input : vec128)
352     -- moods inline
353   return vec128 is
354     variable output : vec128;
355     function flip(input:vec32) return vec32 is
356     -- moods inline
357     begin
358       return input(7 downto 0) & input(15 downto 8)
359           & input(23 downto 16) & input(31 downto 24);
360     end;
361   begin
362     return flip(input(127 downto 96)) & flip(input(95 downto 64))
363         & flip(input(63 downto 32)) & flip(input(31 downto 0));
364   end;
365
366   -----------------------------------------------------------------
367   function aes_keyexpansion(key : vec128)
368     -- moods inline
369   return vec1408 is
370     variable iok : vec128;
371     variable er0,er1,er2,er3,er4,er5,er6,er7,er8,er9:vec128;
372     --variable zero: vec128;
373     --variable expandedkeys: vec1408;
374
375     function exp_round(input : vec128; round: int9) return vec128 is
376     -- moods inline
377       variable r1,r2,r3,r4,r5: vec32;
378     begin
379     r1 := sbox(input(7 downto 0)) &
380           sbox(input(31 downto 24)) &
381           sbox(input(23 downto 16)) &
382           (sbox(input(15 downto 8)) xor rcon(round));
383
384     r2 := input(127 downto 96) xor r1;
385     r3 := input(95 downto 64) xor r2;
```

```
386        r4 := input(63 downto 32) xor r3;
387        r5 := input(31 downto 0) xor r4;
388        return r2 & r3 & r4 & r5;
389      end;
390
391    begin
392    -- first round
393    iok := input_output(key);
394    -- other rounds
395    er9 := exp_round(iok,9);
396    er8 := exp_round(er9,8);
397    er7 := exp_round(er8,7);
398    er6 := exp_round(er7,6);
399    er5 := exp_round(er6,5);
400    er4 := exp_round(er5,4);
401    er3 := exp_round(er4,3);
402    er2 := exp_round(er3,2);
403    er1 := exp_round(er2,1);
404    er0 := exp_round(er1,0);
405
406    return (iok & er9 & er8 & er7 & er6 & er5 & er4 & er3 & er2 & er1 & er0);
407    end;
408
409    ---------------------------------------------------------------
410
411    function aes_core (signal plaintext : vec128; signal keytext : vec128;
             signal encrypt : std_logic)
412        -- moods inline
413    return vec128 is
414      variable rk0 : vec128;
415      variable ciphertext, expkey : vec128;
416      variable ct1, ct2,ct3,ct4,ct5,ct6,ct7,ct8: vec128;
417      variable expandedkeys : vec1408;
418    begin
419    -- expanded key schedule
420    expandedkeys := aes_keyexpansion(keytext);
421
422    -- round 0
423    ct1 := input_output(plaintext) xor expandedkeys(1407 downto 1280);
424
425    -- round 1 to nr-1
426    --for i in 1 to nr-1 loop
427    for i in 1 to 9 loop
428      ct2 := subbytes(ct1);
429      ct3 := shiftrows(ct2);
430      ct4 := mixcolumns(ct3);
431      case(i) is
432       when 1 => expkey := expandedkeys(1279 downto 1152);
433       when 2 => expkey := expandedkeys(1151 downto 1024);
434       when 3 => expkey := expandedkeys(1023 downto 896);
435       when 4 => expkey := expandedkeys(895  downto 768);
436       when 5 => expkey := expandedkeys(767  downto 640);
437       when 6 => expkey := expandedkeys(639  downto 512);
438       when 7 => expkey := expandedkeys(511  downto 384);
439       when 8 => expkey := expandedkeys(383  downto 256);
```

```
440          when 9 => expkey := expandedkeys(255  downto 128);
441          when others => null;
442        end case;
443        ct1 := ct4 xor expkey;
444      end loop;
445
446      -- final round nr=10
447      ct5 := subbytes(ct1);
448      ct6 := shiftrows(ct5);
449      ct7 := ct6 xor expandedkeys(1407-128*nr downto 1280-128*nr);
450
451      ciphertext := input_output(ct7);
452
453      return ciphertext;
454    end;
455
456
457
458    ------------------------------------------------------------------
459    function rcon (input : int9)
460      -- moods inline
461    return vec8 is
462      type rcont_t is array(0 to 9) of vec8;
463      constant table_rcon: rcont_t := (
464      -- moods rom
465      x"36", x"1b", x"80", x"40", x"20", x"10", x"08", x"04", x"02", x"01");
466    begin
467      return table_rcon(input);
468    end;
469
470    ------------------------------------------------------------------
471    function sbox (pt : vec8)
472      -- moods inline
473    return vec8 is
474      variable ct : vec8;
475      type table256 is array(0 to 255) of vec8;
476      constant pt_256 : table256 := (
477      -- moods rom
478      x"63", x"7c", x"77", x"7b", x"f2", x"6b", x"6f", x"c5",
479       x"30", x"01", x"67", x"2b", x"fe", x"d7", x"ab", x"76",
480       x"ca", x"82", x"c9", x"7d", x"fa", x"59", x"47", x"f0",
481       x"ad", x"d4", x"a2", x"af", x"9c", x"a4", x"72", x"c0",
482       x"b7", x"fd", x"93", x"26", x"36", x"3f", x"f7", x"cc",
483       x"34", x"a5", x"e5", x"f1", x"71", x"d8", x"31", x"15",
484       x"04", x"c7", x"23", x"c3", x"18", x"96", x"05", x"9a",
485       x"07", x"12", x"80", x"e2", x"eb", x"27", x"b2", x"75",
486       x"09", x"83", x"2c", x"1a", x"1b", x"6e", x"5a", x"a0",
487       x"52", x"3b", x"d6", x"b3", x"29", x"e3", x"2f", x"84",
488       x"53", x"d1", x"00", x"ed", x"20", x"fc", x"b1", x"5b",
489       x"6a", x"cb", x"be", x"39", x"4a", x"4c", x"58", x"cf",
490       x"d0", x"ef", x"aa", x"fb", x"43", x"4d", x"33", x"85",
491       x"45", x"f9", x"02", x"7f", x"50", x"3c", x"9f", x"a8",
492       x"51", x"a3", x"40", x"8f", x"92", x"9d", x"38", x"f5",
493       x"bc", x"b6", x"da", x"21", x"10", x"ff", x"f3", x"d2",
494       x"cd", x"0c", x"13", x"ec", x"5f", x"97", x"44", x"17",
```

```
495          x"c4", x"a7", x"7e", x"3d", x"64", x"5d", x"19", x"73",
496          x"60", x"81", x"4f", x"dc", x"22", x"2a", x"90", x"88",
497          x"46", x"ee", x"b8", x"14", x"de", x"5e", x"0b", x"db",
498          x"e0", x"32", x"3a", x"0a", x"49", x"06", x"24", x"5c",
499          x"c2", x"d3", x"ac", x"62", x"91", x"95", x"e4", x"79",
500          x"e7", x"c8", x"37", x"6d", x"8d", x"d5", x"4e", x"a9",
501          x"6c", x"56", x"f4", x"ea", x"65", x"7a", x"ae", x"08",
502          x"ba", x"78", x"25", x"2e", x"1c", x"a6", x"b4", x"c6",
503          x"e8", x"dd", x"74", x"1f", x"4b", x"bd", x"8b", x"8a",
504          x"70", x"3e", x"b5", x"66", x"48", x"03", x"f6", x"0e",
505          x"61", x"35", x"57", x"b9", x"86", x"c1", x"1d", x"9e",
506          x"e1", x"f8", x"98", x"11", x"69", x"d9", x"8e", x"94",
507          x"9b", x"1e", x"87", x"e9", x"ce", x"55", x"28", x"df",
508          x"8c", x"a1", x"89", x"0d", x"bf", x"e6", x"42", x"68",
509          x"41", x"99", x"2d", x"0f", x"b0", x"54", x"bb", x"16");
510        begin
511          ct := pt_256(to_integer(unsigned(pt)));
512          return ct;
513        end;
514
515      _____
516      end;
```

After the functions and top level entity have been defined, we can implement a test bench that applies a set of test data to the inputs and verifies that the correct output has been obtained. Notice that we use the assertion technique to identify correct operation.

```
1      library ieee;
2      use ieee.std_logic_1164.all;
3      entity testAES is
4      end;
5
6
7      library ieee;
8      use ieee.std_logic_1164.all;
9      use work.aes_functions.all;
10     architecture behavior of testAES is
11
12       component aes
13         port(
14           plaintext : in std_logic_vector(127 downto 0);
15           keytext : in std_logic_vector(127 downto 0);
16           encrypt : in std_logic;
17           go : in std_logic;
18           ciphertext : out std_logic_vector(127 downto 0);
19           done : out std_logic
20         );
21       end component;
22
23       for all : aes use entity work.aes;
24
25       signal plaintext : std_logic_vector(127 downto 0);
26       signal keytext : std_logic_vector(127 downto 0);
27       signal encrypt : std_logic;
28       signal go : std_logic := '0';
```

```
29      signal ciphertext : std_logic_vector(127 downto 0);
30      signal done : std_logic;
31      signal ok : std_logic := '0';
32   begin
33      plaintext <=  X"0000000000000000000000000000000", X"3243
            f6a8885a308d313198a2e0370734" after 50 ns ;
34      keytext <=  X"00000000000000000000000000000000", X"2
            b7e151628aed2a6abf7158809cf4f3c" after 100 ns;
35      process (ciphertext)
36           variable ct : std_logic_vector(127 downto 0) := X"3925841
                 d02dc09fbdc118597196a0b32";
37      begin
38         assert ct = ciphertext
39         report "Test vectors do not match"
40       severity note;
41         assert not (ct = ciphertext)
42         report "Test vectors Matched"
43         severity note;
44      end process;
45
46
47      process
48      begin
49        go <= not go after 20 ns;
50      end process;
51
52      DUT : aes port map (plaintext, keytext, encrypt, go, ciphertext, done);
53   end;
```

10.5 Summary

This chapter shows how standard block ciphers can be implemented in VHDL and Verilog using DES as an example. AES has been developed further using VHDL, but can use the same principles in Verilog. Both of these algorithms are in common usage today and in operational hardware. There are numerous other methods, as security requires a constant evolution of encryption techniques and no doubt more robust and secure methods will emerge that require implementation in VHDL and/or Verilog in the future.

CHAPTER 11

Memory

11.1 Introduction

There are very often two ways to use memory on modern FPGAs. Either there are memory blocks on board the FPGA (or on the development board) or you wish to make your own memory blocks for storage using the flip flops on the FPGA logic. Either way, it is important to realize that there are significant differences between dedicated high density Dynamic RAM (DRAM) and Synchronous Dynamic Random Access Memory (SDRAM) and flip-flop based Static RAM (SRAM).

If we consider SDRAM, the key aspects of this type of memory to consider are:

1. This type of DRAM relies on transistor capacitance on gates to store data.
2. DRAM is much more compact than SRAM (Static RAM).
3. DRAM cannot be synthesized; you need a separate DRAM chip.
4. SDRAM requires a synchronization clock that is consistent with the rest of the hardware system (it is designed to operate with microprocessors).
5. DRAM data must be refreshed as it is stored charge and decays after a certain time.
6. DRAM is slower than SRAM.

Static RAM can be considered in a similar way to a ROM chip and it also has (differing) key aspects of behavior to consider:

1. Memory cells are based on standard latches.
2. SRAM is fast.
3. SRAM is less compact than DRAM (or SDRAM).
4. SRAM can be synthesized on an FPGA so is ideal for small, fast registers or memory blocks.

Static RAM is essentially asynchronous, but can be modified to behave synchronously (as SDRAM is the synchronous equivalent of DRAM), and this is often called Synchronous RAM.

Flash memory is useful to consider at this point, even though its operation is fundamentally different from the memory types considered thus far, simply because it is easy to use and is commonly available on many FPGA development boards. Flash memory is essentially a form of EEPROM (electrically programmable ROM) that can be used as a form of persistent RAM. Why persistent? In flash memory, the device memory is retained even when the power is removed, so it is often used as a form of ROM, which makes it an interesting memory to use on FPGA systems as it could be used to store the FPGA program, but also used as a RAM storage (dynamically) for current data.

11.2 Modeling Memory in HDLs

Great care must be exercised when modeling memory in HDLs. As some memory cannot be synthesized, if a model is used, it must reflect the correct physical behavior of the real device if it is off chip. This particularly applies to access times and timing violation conditions. If the timing is violated, then the data may be at best suspect and at worst totally useless. The designer can find themselves in the invidious position of having a simulation model that works perfectly, and real hardware that is completely nonfunctional. In this chapter, no physical delays have been implemented in the models, and these must be added if the models are to be used in a realistic system.

11.3 Read Only Memory

Read only memory (ROM) is essentially a set of predefined data values in a storage register. The memory has two definitions: first, the number of storage areas and second, the number of bits. For example, if the memory has 16 storage areas and 8 bits each, the memory is defined as a 16×8 ROM. The basic ROM has one input, the definition of the address to be accessed, and one output, which is a logic vector which is where the data will be put. Consider as an example the entity for a simple behavioral ROM model in VHDL:

```
1    entity ROM16x8 is
2      port (
3        address : in integer range 0 TO 15;
4        dout : out std_logic_vector (7 downto 0)
5        );
6    end entity ROM16x8;}
```

As can be seen, the address has been defined as an integer, but the range has been restricted to the range of the ROM. The architecture of the ROM is defined as a fixed array of elements that can be accessed directly. Therefore, an example ROM with a set of example data elements could be defined as follows:

```
1    architecture example of rom16x8 is
2      type romdata is array (0 to 15)
3    of std_logic_vector( 7 downto 0);
4      constant romvals : romdata := (
5        "00000000",
```

```
 6          "01010011",
 7          "01110010",
 8          "01101100",
 9          "01110101",
10          "11010111",
11          "11011111",
12          "00111110",
13          "11101100",
14          "10000110",
15          "11111001",
16          "00111001",
17          "01010101",
18          "11110111",
19          "10111111",
20          "11101101");
21       begin
22         data <= romvals(address);
23       end architecture example;
```

If we wish to use this in an example, we first need to declare the ROM in a VHDL testbench and then specify the address using an integer signal. A sample testbench is given here:

```
 1       library ieee;
 2       use ieee.std_logic_1164.all;
 3
 4       entity ROM16x8_TB is
 5       end entity ROM16x8_TB;
 6
 7       architecture TB of ROM16x8_TB is
 8             signal address : integer := 0;
 9             signal data : std_logic_vector ( 7 downto 0 );
10       begin
11             ROM16x8: entity work.ROM16x8(example)
12       port map ( address, data );
13       end architecture TB;
```

Notice that the IEEE library, std_logic_1164, is required for the std_logic_vector type and the value of the data will depend on the address chosen.

We can implement a very similar type of fixed ROM using a case statement in Verilog and the listing for the equivalent 16 × 8 ROM is provided as follows:

```
 1       module rom16x8 (
 2         address , // Address input
 3         data       , // Data output
 4       );
 5       input [3:0] address;
 6       output [7:0] data;
 7
 8       reg [7:0] data ;
 9
10       always @ (address)
11       begin
12         case (address)
13           0 : data = 8'b00000000;
```

```
14        1 : data = 8'b01010011;
15        2 : data = 8'b01110010;
16        3 : data = 8'b01101100;
17        4 : data = 8'b01110101;
18        5 : data = 8'b11010111;
19        6 : data = 8'b11011111;
20        7 : data = 8'b00111110;
21        8 : data = 8'b11101100;
22        9 : data = 8'b10000110;
23        10 : data = 8'b11111001;
24        11 : data = 8'b00111001;
25        12 : data = 8'b01010101;
26        13 : data = 8'b11110111;
27        14 : data = 8'b10111111;
28        15 : data = 8'b11101101;
29     endcase
30   end
31
32   endmodule
```

11.4 Random Access Memory

A dynamic random access memory (RAM) block has a two-dimensional structure of memory that is divided into a grid structure that can be accessed by a row address and column address (obviously this is one way of carrying this out, but it is often the approach required for dynamic RAM). Note that this is asynchronous and therefore has no clock. The implication of being asynchronous is that great care must be taken with the timing of the memory access to ensure data integrity throughout the transfer process.

The VHDL model has a single address input and two control signals *row* and *col* used to load the row and column address, respectively. There is also a *rw* signal that is defined as being write when high and read when low. Finally, the data is put onto the *data* signal, which is defined as an *inout* (bidirectional) signal. The resulting model is given in the VHDL following. In this example, the number of rows is 2^8 and the number of columns also 2^8. This gives a total data storage with 16 bits of 1 MBit.

```
1    library ieee;
2    use ieee.std_logic_1164.all;
3    use ieee.numeric_std.all;
4
5    entity dram1mb is
6      port (
7        address : in integer range 0 to 2**8-1; -- Row Address
8        row : in std_logic; -- Row Select
9        col : in std_logic; -- Column Select
10       rw : std_logic; -- Read/Write
11       data : inout std_logic_vector (15 downto 0) -- Data
12      );
13   end entity dram1mb;
14
```

```
15    architecture behav of dram1mb is
16    begin
17    process (row, col, rw) is
18      type dram is array (0 to 2**16 - 1) of std_logic_vector(15 downto 0);
19      variable radd: INTEGER range 0 to 2**8 - 1;
20      variable madd: INTEGER range 0 to 2**16 - 1;
21      variable memory: dram;
22    begin
23      data <= (others => 'Z');
24      if falling_edge(row) then
25        radd := address;
26      elsif falling_edge(col) then
27        madd :=radd*2**8 +address;
28        if row = '0' and rw = '0' then
29          memory(madd) := data;
30        end if;
31      elsif col = '0' and row = '0' and rw = '1' then
32        data <= memory(madd);
33      end if;
34    end process;
35    end architecture behav;
```

Using this model a simple testbench can be used to read in a data value to an address, then another value to another address and then the original value read back. The test bench to achieve this is given in the VHDL.

```
1     library ieee;
2     use ieee.std_logic_1164.all;
3
4     entity testram is
5     end entity testram;
6
7     architecture test of testram is
8         signal address : integer range 0 to 2**8-1 := 0;
9         signal rw : std_logic;
10        signal c : std_logic;
11        signal r : std_logic;
12        signal data : std_logic_vector ( 15 downto 0 );
13    begin
14
15        dram: entity work.dram1mb(behav)
16    port map ( address, r, c, rw, data );
17        address <= 23 after 0 ns, 47 after 30 ns, 23 after 90 ns;
18        rw <= '0' after 0 ns, '1' after 90 ns;
19        c <=          '1' after 0 ns, '0' after 20 ns,
20    '1' after 50 ns, '0' after 70 ns,
21                             '1' after 90 ns, '0' after 100 ns;
22        r <=              '1' after 0 ns, '0' after 10 ns,
23    '1' after 40 ns, '0' after 60 ns,
24                             '1' after 80 ns, '0' after 100 ns;
25        data <=          X"1234" after 0 ns, X"5678" after 40 ns;
26
27    end architecture test;
```

The results of testing this model can be seen in the waveform plot in Figure 11.1, which shows the correct behavior of the address, data, and control lines.

The Verilog model has a single address input and two control signals *row* and *col* used to load the row and column address, respectively. There is also a *rw* signal that is defined as being write when high and read when low. Finally, the data is put onto the *datain* signal, which is defined as an *input* signal and read back from the *dataout* signal which is defined as a register. The resulting model is given in the Verilog following. In this example, the number of rows is 2^8 and the number of columns also 2^8. This gives a total data storage with 16 bits of 1 MBit.

```
1    module dram1mb ( address, row, col, rw, datain, dataout );
2
3        input [7:0] address; // Address
4        input row;              // Row Selector
5        input col;              // Column Selector
6        input rw;              // Read/Write
7        input [15:0] datain ;    // Data In
8        output [15:0] dataout ;  // Data Out
9
10       wire [15:0] datain;
11       reg [15:0] dataout ;  // Data defined as a register
12
13       reg [7:0] radd;      // Row Address
14       reg [15:0] madd;     // Overall memory Address
15
16       // define the memory array
17       reg [15:0] memory [0:255]; // the memory array is 16 bits wide (data) and
            2**8 -1 deep (address)
18
19       always @ (negedge row)
20       begin
```

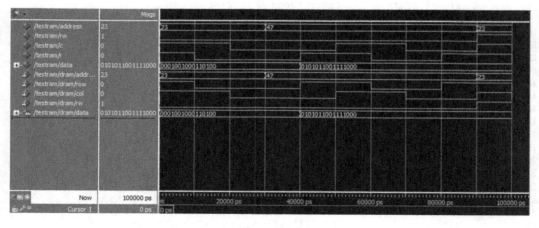

Figure 11.1
Basic VHDL RAM simulation.

```
21        radd = address;
22    end
23
24    always @ (negedge col)
25    begin
26      madd = radd*2**8 + address;
27          if(!rw) begin
28            memory[madd] <= datain;
29          end
30          else begin
31            dataout <= memory[madd];
32          end
33      end
34    endmodule
```

Using this model, a simple testbench can be used to read in a data value to an address, then another value to another address and then the original value read back. The test bench to achieve this is given in the Verilog following. Note as stated earlier that this is an asynchronous model, which means that the *rw* signal must be defined prior to the *col* control signal going low (and it is very important to not make this coincident with the *rw* signal, otherwise a race condition would arise).

```
1
2     module dram1mb_tb();
3     // declare the counter signals
4     reg row;
5     reg col;
6     reg rw;
7     reg [7:0] address;
8     wire [15:0] dataout;
9
10    reg [15:0] datain;
11
12    // Set up the initial variables and reset
13    initial begin
14      $display ("time\t row col rw address data data_set");
15      $monitor ("%g\t %b %b %b %d %d %d",
16      $time, row, col, rw, address, datain, dataout);
17      row = 1;    // initialize the row to 1
18      col = 1;    // set the col to 1
19      rw = 1;     // set the rw to 1
20
21      #1 address = 0; // set the row to 1
22      #1 datain = 23;
23      #1 row = 0; // set the row to 0
24      #1 row = 1; // set the row to 1
25      #1 address = 1; // set the row to 1
26      #1 rw = 0; // set the rw to 0
27      #1 col = 0; // set the col to 0
28      #1 col = 1; // set the col to 1
29      #1 rw = 1; // set the rw to 1
30
31      // This should have written the data of 23 into location 0:1
32
```

```
33
34    #1 address = 0; // set the row to 1
35    #1 datain = 47; // set the data to 47
36    #1 row = 0; // set the row to 0
37    #1 row = 1; // set the row to 1
38    #1 address = 2; // set the row to 2
39    #1 rw = 0; // set the rw to 0
40    #1 col = 0; // set the col to 0
41    #1 col = 1; // set the col to 1
42    #1 rw = 1; // set the rw to 1
43
44    // This should have written the data of 47 into location 0:2
45
46    #1 address = 0; // set the row to 1
47    #1 row = 0; // set the row to 0
48    #1 row = 1; // set the row to 1
49    #1 address = 1; // set the row to 1
50    #1 col = 0; // set the col to 0
51    #1 col = 1; // set the col to 1
52
53    #100 $finish;    // Finish the simulation
54    end
55
56    dram1mb RAM ( address, row, col, rw, datain, dataout);
57
58    endmodule
```

The results of testing this model can be seen in the waveform plot in Figure 11.2 which shows the correct behavior of the address, data, and control lines.

This model is different from the VHDL in an important aspect: that the memory has a separate data input and output port (defined as an input and output, respectively). This is not what was defined in the VHDL model, which used a single *inout* port for the data. Therefore, how can

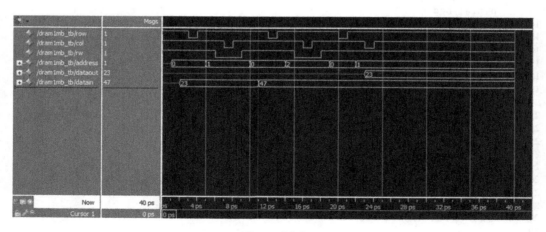

Figure 11.2
Basic Verilog RAM simulation.

this be done using Verilog? We have two basic types in verilog, *wire* for direct connections and *reg* for registers. If we use an *inout* type for the port, then a register cannot be used, but also a wire does not hold the value, so how can we reconcile this in a RAM model? The answer is to declare an internal register to hold the output value of the data port, and then, depending on the value of the *rw* command signal, define the data port as having the value defined by the output register (in this model *dataout*) or making it a high impedance state (tri-state). This port can then be driven externally, as is shown in the test bench.

```
1    module dram1mbio ( address, row, col, rw, data );
2     input [7:0] address; // Address
3     input row;           // Row Selector
4     input col;           // Column Selector
5     input rw;            // Read/Write
6     inout [15:0] data ;  // Data In
7
8     wire [15:0] data;
9     reg [15:0] dataout ;   // Data defined as a register
10
11    reg [7:0] radd;    // Row Address
12    reg [15:0] madd;   // Overall memory Address
13
14    // define the memory array
15    reg [15:0] memory [0:255]; // the memory array is 16 bits wide (data) and
            2**8 -1 deep (address)
16    assign data = (rw) ? dataout : 16'hz;
17    always @ (negedge row)
18    begin
19      radd = address;
20    end
21
22    always @ (negedge col)
23    begin
24      madd = radd*2**8 + address;
25          if(!rw) begin
26            memory[madd] <= data;
27          end
28          else begin
29            dataout <= memory[madd];
30          end
31    end
32    endmodule
```

Using this model a simple testbench can be used to read in a data value to an address, then another value to another address, and then the original value read back. The test bench to achieve this is given in the Verilog following. Note as stated earlier that this is an asynchronous model, which means that the *rw* signal must be defined prior to the *col* control signal going low (and it is very important to not make this coincident with the *rw* signal, otherwise a race condition would arise). In the test bench, it can be seen that the same tri-state technique is required as in the RAM model itself, allowing the single port to be used as an input to read the data into the memory or as a register to write the requested data value back out to the data bus.

```
1
2    module dram1mbio_tb();
3    // declare the counter signals
4    reg row;
5    reg col;
6    reg rw;
7    reg [7:0] address;
8    wire [15:0] data;
9
10   reg [15:0] datain;
11
12   assign data = (!rw) ? datain : 16'hz;
13
14   // Set up the initial variables and reset
15   initial begin
16     $display ("time\t row col rw address data data_set");
17     $monitor ("%g\t %b %b %b %d %d %d",
18     $time, row, col, rw, address, datain, data);
19     col = 1;          // set the col to 1
20     rw = 1;           // set the rw to 1
21
22     #1 address = 0;   // set the row to 1
23     #1 datain = 23;
24     #1 row = 0;    // set the row to 0
25     #1 row = 1;    // set the row to 1
26     #1 address = 1;  // set the row to 1
27     #1 rw = 0;    // set the rw to 0
28     #1 col = 0;   // set the col to 0
29     #1 col = 1;   // set the col to 1
30     #1 rw = 1;    // set the rw to 1
31
32     // This should have written the data of 23 into location 0:1
33
34
35     #1 address = 0;    // set the row to 1
36     #1 datain = 47;    // set the data to 47
37     #1 row = 0;     // set the row to 0
38     #1 row = 1;      // set the row to 1
39     #1 address = 2;  // set the row to 2
40     #1 rw = 0;      // set the rw to 0
41     #1 col = 0;    // set the col to 0
42     #1 col = 1;    // set the col to 1
43     #1 rw = 1;     // set the rw to 1
44
45     // This should have written the data of 47 into location 0:2
46
47     #1 address = 0;  // set the row to 1
48     #1 row = 0;   // set the row to 0
49     #1 row = 1;    // set the row to 1
50     #1 address = 1;   // set the row to 1
51     #1 col = 0;   // set the col to 0
52     #1 col = 1;   // set the col to 1
53
54     #100 $finish;        // Finish the simulation
55   end
```

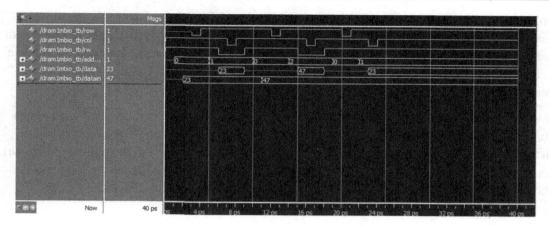

Figure 11.3
Basic RAM simulation with a common data port.

```
56
57    dram1mbio RAM ( address, row, col, rw, data);
58
59    endmodule
```

The results of testing this model can be seen in the waveform plot in Figure 11.3 which shows the correct behavior of the address, data, and control lines.

It is important to note that the RAM model in both VHDL and Verilog cases does not model any of the actual delays that would appear in practice and if this is important to the functionality of the design, then it MUST be added to the model. Also, note that this could be obtained from a data sheet for DRAM (dynamic RAM); however, the delays would have an uncertainty attached to them.

11.5 Synchronous RAM

In the preceding chapter, we observed how the memory is accessed asynchronously, whereas synchronous RAM (SRAM) requires a clock. In most practical designs, the RAM will be implemented off-chip as a separate memory device, but sometimes it is useful to define a small block of RAM on the FPGA for fast access or local storage close to the hardware device that requires frequent access to a relatively small memory block.

The usual design constraints apply to memory, more so than some other possible functions, as the use of flip-flops to store data without using much of the other logic in a look-up table (LUT) is area intensive. The trade-off, as ever with FPGA design, is whether the potential for improved performance and speed using on-board RAM outweighs the increased area required as a result.

From the design perspective, the synchronous RAM model is very similar to the previously demonstrated basic asynchronous RAM model. The only difference is that, instead of the data being available immediately on the address being applied (or after some short delay), the data in a synchronous RAM is only accessed when the clock edge occurs (rising or falling edge depending on the design required).

If we consider the VHDL for the SRAM, we can see that for a memory size of 2^m and a data bus of 2^n, the following model is required. The VHDL model has two parameters, m and n. In the default case, the value of m as 16 provides 64k address words and the number of bits (n) set to 16 gives a total of 1M bits in the RAM. Obviously this could be made any size, but this shows the type of calculation required to obtain the specified memory blocks.

```
1    library ieee;
2    use ieee.std_logic_1164.all;
3    use ieee.numeric_std.all;
4
5    entity sram is
6      generic (
7        m : natural := 16;
8        n : natural := 16
9      );
10     port (
11       clk : in std_logic;
12         addr : in std_logic_vector (m-1 downto 0);
13         wr : in std_logic;
14       d : in std_logic_vector (n-1 downto 0);
15         q : out std_logic_vector (n-1 downto 0)
16     );
17   end entity sram;
18
19   architecture dualport of sram is
20     type sramdata is array (0 to 2**m-1) of std_logic_vector (n-1 downto 0);
21     signal memory : sramdata;
22   begin
23     process (clk ) is
24     begin
25       if rising_edge(clk) then
26         if wr = '0' then
27           memory(to_integer(unsigned(addr))) <= d;
28         else
29           q <= memory(to_integer(unsigned(addr)));
30         end if;
31       end if;
32     end process;
33   end architecture dualport;
```

Notice that there are two control signals, the clock, *clk*, and the write enable, *wr*. We could make the memory synchronous write, synchronous read, or a more complex port structure, but in this case, we will show the operation as being synchronous read and write, on the rising edge of the clock. Also, the convention we will use is for the write enable state to be active when wr is low.

There are several interesting aspects to this model that are worth considering. The first is the access of the memory. If we define the address as a std_logic_vector type in VHDL, then we can't simply use this value to access a specific element of an array. This requires an integer argument. We also cannot simply cast a std_logic_vector type directly to an integer. The first thing we must do is convert the std_logic_vector type to an unsigned number. This is a "halfway house" from std_logic_vector to integer, in that we can use the variable as a number, but it is limited to the same bit resolution as the original std_logic_vector. In this case, clearly this is not an issue as we do not want the address to be larger than the memory, otherwise errors will result. The final step is to convert the unsigned type to an integer. This is accomplished using the to_integer function and is the final step to convert the address into the integer form required to access the individual element of the array. As a consequence of using these numeric functions, we need to also include the IEEE standard numeric library in the header of the model as shown:

```
1    library ieee;
2    use ieee.std_logic_1164.all;
3    use ieee.numeric_std.all;
```

It is also worth noting that the read and write functions are mutually exclusive, in that you cannot read from the memory and write to it at the same time. This ensures the integrity of the data. Also note that the read and write functions are both clocked and so the memory is both read and write synchronous.

We can test this model by using a test bench similar to that used for the previous RAM models as given in the following listing:

```
1    library ieee;
2    use ieee.std_logic_1164.all;
3    use ieee.numeric_std.all;
4
5    entity testram is
6    end entity testram;
7
8    architecture test of testram is
9        signal address : std_logic_vector ( 7 downto 0 );
10       signal clk : std_logic;
11       signal wr : std_logic;
12       signal d : std_logic_vector ( 15 downto 0 );
13       signal q : std_logic_vector ( 15 downto 0 );
14       constant period : time := 5 ns;
15    begin
16
17   sram: entity work.sram(dualport) generic map ( 8, 16) port map( clk, address,
         wr, d, q ) ;
18
19   address <= "00000001" after 0 ns, "00000010" after 30 ns, "00000001" after 90
         ns;
20   wr <=  '0' after 0 ns, '1' after 90 ns;
21   d <=  X"1234" after 0 ns, X"5678" after 40 ns;
22
```

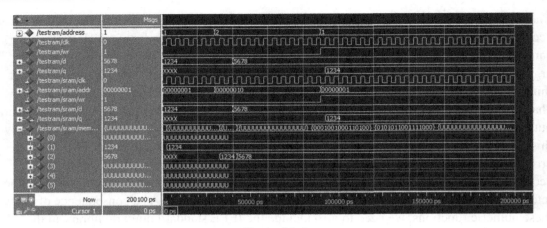

Figure 11.4
Basic VHDL SRAM Simulation.

```
23
24      -- Clock process definitions( clock with 50% duty cycle is generated here.
25      clk_process :process
26      begin
27        clk <= '0';
28        wait for period/2;
29        clk <= '1';
30        wait for period/2;
31      end process;
32
33      end architecture test;
```

The results of testing this model can be seen in the waveform diagram in Figure 11.4, which shows the correct behavior of the address, data, and control lines.

Implementing the SRAM in Verilog is similar to the dual port RAM model earlier in this chapter, with an address, data input, and output ports, but importantly this model is now synchronous and so the management of the data to and from the memory is controlled by the clock signal *clk*.

```
1       module sram_verilog ( address, clk, rw, d, q );
2         parameter m = 8; // Address Bus Width
3         parameter n = 16; // Data Bus Width
4         input [m-1:0] address; // Address
5         input clk;            // Clock
6         input rw;             // Read/Write
7         input [n-1:0] d ;     // Data In
8         output [n-1:0] q ;    // Data Out
9
10        wire [n-1:0] d;
11        reg [n-1:0] q ;       // Data defined as a register
12
```

```
13
14          // define the memory array
15          reg [n-1:0] memory [0:2**m-1]; // the memory array is n bits wide (data) and
                  2**m -1 deep (address)
16
17          always @ (posedge clk)
18          begin
19                  if(!rw) begin
20                    memory[address] <= d;
21                  end
22                  else begin
23                    q <= memory[address];
24                  end
25          end
26      endmodule
```

The test bench of the model is almost identical to the previous memory test benches; however, the clock is defined in the test bench, and as a result the variables can be seen to be changing at a rate dependent on the clock (in this case the clock changes every 5 'ticks' and the other signals are therefore defined to change every 10 'ticks').

```
1
2       module sram_verilog_tb();
3       // declare the counter signals
4       parameter m = 8;
5       parameter n = 16;
6       reg rw;
7       reg clk;
8       reg [m-1:0] address;
9       wire [n-1:0] dataout;
10
11      reg [n-1:0] datain;
12
13      // Set up the initial variables and reset
14      initial begin
15          $display ("time\t rw address data data_set");
16          $monitor ("%g\t %b %d %d %d",
17          $time, rw, address, datain, dataout);
18          rw = 1;     // set the rw to 1
19          clk = 0;    // INit the clk to 0
20
21          #10 address = 0; // set the row to 1
22          #10 datain = 23;
23          #10 rw = 0; // set the rw to 0
24          #10 address = 1; // set the row to 1
25
26          #10 rw = 1; // set the rw to 1
27
28          // This should have written the data of 23 into location 0:1
29
30
31          #10 address = 0; // set the row to 1
32          #10 datain = 47; // set the data to 47
33          #10 address = 2; // set the row to 2
34          #10 rw = 0; // set the rw to 0
```

```
35    #10 rw = 1; // set the rw to 1
36
37    // This should have written the data of 47 into location 0:2
38
39    #10 address = 0; // set the row to 1
40    #10 address = 1; // set the row to 1
41
42
43    #100 $finish; // Finish the simulation
44  end
45
46  // Clock
47  always begin
48    #5 clk = ~clk; // Invert the clock every 5 time ticks
49  end
50
51  sram_verilog RAM ( address, clk, rw, datain, dataout);
52
53  endmodule
```

The results of testing this model can be seen in the waveform diagram in Figure 11.5, which shows the correct behavior of the address, data, and control lines.

11.6 Flash Memory

As has been discussed previously, flash memory is essentially a form of EEPROM (Electrically Erasable and Programmable Read Only Memory). This is slightly different from a standard RAM where the address is given to the memory and depending on the R/W signals, the data is read or written, respectively. A typical set of interface pins for a flash memory consists of the following elements:

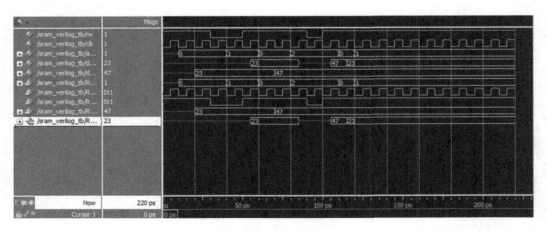

Figure 11.5
Basic Verilog SRAM Simulation.

Pin	Function	Active State
CLE	Command Latch	H, activated on rising_edge(WE)
ALE	Address Latch	H, activated on rising_edge(WE)
CE	Chip Enable	L
RE	Read Enable	Falling_edge(RE)
WE	Write Enable	Rising_edge(WE)
WP	Write protect	L
busy	Ready/Busy	L = busy, H = ready

In addition to these control signals there is of course an address bus and a data bus. To implement this we can use a similar entity to that for a standard RAM block in VHDL:

```
1    entity flash is
2      generic (
3        a : natural := 10;
4        d : natural := 8
5      );
6      port (
7        clk : in std_logic;
8        addr : in std_logic_vector(a-1 downto 0);
9        data : inout std_logic_vector (d-1 downto 0);
10       cle : in std_logic;
11       ale : in std_logic;
12       ce : in std_logic;
13       re : in std_logic;
14       we : in std_logic;
15       wp : in std_logic;
16       busy : out std_logic;
17     );
18   end entity flash;
```

In most cases we won't need to model the flash memory itself, but rather we need to interface to it, so the entity for a flash interface controller could be as follows:

```
1    entity flashif is
2      port (
3        clk : in std_logic;
4        read : in std_logic;
5        en : in std_logic;
6        cle : out std_logic;
7        ale : out std_logic;
8        ce : out std_logic;
9        re : out std_logic;
10       we : out std_logic;
11       wp : out std_logic;
12       busy : in std_logic;
13     );
14   end entity flashif;
```

A typical architecture for this device could be as follows:

```
1    architecture basic of flashif is
2    begin
3      process (clk) is
4        if busy =   1  then
5          if rising_edge(clk) then
6            ce <= en;
7            ale <=   1  ;
8            cle <=   1  ;
9            if read =    0  then
10             we <=   1   ;
11             re <=   1   ;
12           else
13             we <=   0   ;
14             re <=   0   ;
15           end if;
16           if prog =    0  then
17             wp <=   0   ;
18           else
19             wp <=   1   ;
20           end if;
21         end if;
22       end if;
23     end process;
24   end architecture basic;
```

This is a basic outline for a flash controller and this will obviously change from device to device.

11.7 Summary

This chapter has introduced the important memory types of ROM, asynchronous RAM, FLASH memory, and synchronous RAM. It is important to remember that in most cases, large memory blocks will be contained off chip and so it may be necessary to use these models purely for simulation rather than synthesis, but that it is possible to use RAM sparingly on the FPGA itself if absolutely required.

In this case, the trade-off of speed vs. area becomes particularly acute and as such great care must be taken to not make naive decisions about putting large amounts of memory on the FPGA, as this may take up far too much memory to be practical.

CHAPTER 12

PS/2 Mouse Interface

12.1 Introduction

The PS/2 mouse is a standard interface to both computers and also many FPGA development kits. The protocol is a serial one and in this chapter the basics of the protocol will be reviewed as well as a simple VHDL interface code to enable the designer to use a mouse, primarily on a standard FPGA development kit.

12.2 PS/2 Mouse Basics

The origins of the PS/2 mouse are back in the 1980s with the proliferation of the IBM Personal Computer (PC). This had the generic name of a Personal System, hence PS and the second version of this was therefore called the PS/2. The interface technology has remained under that name ever since.

The PS/2 interface is essentially a custom serial interface with one device supported per connector (unlike the modern USB, Universal Serial Bus, which can handle numerous devices on a single port). The data rate is relatively slow (40 kbps) and the device is powered off a 5 V dc supply.

Unlike the USB approach where devices are generally hot swappable, that is, they can be plugged in or unplugged at will), the PS/2 device cannot be removed without a system crash or freeze resulting.

The PS/2 mouse supports communication from the mouse to the host and vice versa, and the supply is provided from the host to the mouse in the form of a 5 V line.

12.3 PS/2 Mouse Commands

The PS/2 mouse has a limited set of commands that are essentially either button press commands or mouse movement commands. The standard mouse has a left, middle, and right button click command, and X and Y movement. The X and Y movements are tracked using counters, where the value is relative to the previous value sent by the mouse, not the absolute position itself.

Design Recipes for FPGAs. http://dx.doi.org/10.1016/B978-0-08-097129-2.00012-X

12.4 PS/2 Mouse Data Packets

The PS/2 mouse sends data in serial packets down a data line and this is synchronous with a clock line also on the mouse interface. Each packet consists of three, 8-bit words where the first word is a configuration word with a set of flags, the second word provides the mouse X movement, and the third word provides the mouse Y movement. The description of the mouse bits are given in the following table:

Bit	Byte 1	Byte 2	Byte 3
7	Y overflow	X Movement 7	Y movement 7
6	X overflow	X Movement 6	Y movement 6
5	Y sign bit	X Movement 5	Y movement 5
4	X sign bit	X Movement 4	Y movement 4
3	Always 1	X Movement 3	Y movement 3
2	Middle Btn	X Movement 2	Y movement 2
1	Right Btn	X Movement 1	Y movement 1
0	Left Btn	X Movement 0	Y movement 0

Each of the movement bytes are defined as 9-bit 2s-complement numbers, where the sign bit is defined in byte 1. The range of movement that can be defined is -255 to $+255$ using this approach.

12.5 PS/2 Operation Modes

The PS/2 mouse operates in four basic modes. On power up the mouse goes into a reset mode and this can also be initiated by a reset command from the host, which is defined as 0xFF. After reset has been completed, the mouse automatically goes into a stream mode in which the data is streamed back from the mouse to the host. These two modes are the most commonly used modes of operation for most applications, but there are two other modes used, which are remote and wrap. These are mostly useful in testing that the interface is operating correctly.

In the reset mode the mouse itself will reset and carry out some basic self checks. The default settings are then defined for the mouse to operate with, which are a sample period of 10 ms, a basic resolution of 4 counts per mm, a 1 to 1 scaling and the data reporting option is disabled.

The mouse sends a device ID of 0x00 to the host to let it know that it is not a keyboard or more complex mouse, just a basic PS/2 mouse.

Once the mouse is running it goes into stream mode and the mouse will send packets to the host at the defined sample rate of activity, such as mouse movement or button presses. The mouse ONLY sends data when activity is present, otherwise it will do nothing.

If the mouse is asked by the host to go into remote mode, then the mouse only sends data when requested by the host, and finally in wrap mode, the mouse sends back every command to the host (apart from the reset and reset wrap commands).

12.6 PS/2 Mouse with Wheel

A mouse that has a wheel is defined as a separate type of device and so it has a different device id 0x03. In this case, after reset, the mouse sends the ID and in the case of a wheel mouse, the data packet is now 4 bytes long and there is an extra byte to provide the wheel movement. This byte only uses the least significant bits in a 2s-complement form and therefore has a range of -8 to $+7$.

12.7 Basic PS/2 Mouse Handler VHDL

The simplest form of the VHDL handler could use the mouse clock signal as the system clock and then monitor the data coming from the mouse. This is shown here:

```
1   library ieee;
2   use ieee.std_logic_1164.all;
3
4   entity psmouse is
5     port (
6       clock : in std_logic;
7       data : in std_logic
8     );
9   end entity psmouse;
10
11  architecture basic of psmouse is
12    signal d : std_logic_vector (23 downto 0);
13    signal byte1 : std_logic_vector (7 downto 0);
14    signal byte2 : std_logic_vector (7 downto 0);
15    signal byte3 : std_logic_vector (7 downto 0);
16    signal index : integer :$=$ 23;
17  begin
18    process(clock) is
19    begin
20      if falling_edge(clock) then
21        d(index) <= data;
22        if index>0 then
23          index <= index−1;
24        else
25          byte1 <= d(23 downto 16);
26          byte2 <= d(15 downto 8);
27          byte3 <= d(7 downto 0);
28          index<=23;
29        end if;
30      end if;
31    end process;
32  end architecture basic;
```

This VHDL is very simple: on each falling edge of the clock the current value of the data is read into the next element of the data array (d) and when the complete 24-bits packet has been read in (and index has counted down to zero) then the three bytes are transcribed from the packet.

12.8 *Modified PS/2 Mouse Handler VHDL*

The trouble with the previous mouse handler is that, although syntactically correct, there could be noise on the mouse clock and data signals leading to an incorrect clocking of the data and so another approach would be to have a much higher frequency signal clock and to monitor the PS/2 clock as if it were a signal. An extra check would be to filter the PS/2 clock so that only if there were a certain number of values the same would the clock be considered to have changed.

```
1    library ieee;
2    use ieee.std_logic_1164.all;
3
4    entity psmouse is
5      port (
6        clk : in std_logic;
7        ps2_clock : in std_logic;
8        data : in std_logic
9      );
10   end entity psmouse;
11
12   architecture basic of psmouse is
13      signal clk_internal : std_logic := 0 ;
14      signal d : std_logic_vector (23 downto 0);
15      signal byte1 : std_logic_vector (7 downto 0);
16      signal byte2 : std_logic_vector (7 downto 0);
17      signal byte3 : std_logic_vector (7 downto 0);
18      signal index : integer := 23;
19   begin
20     process(clock) is
21       high : integer := 0;
22       low : integer := 0;
23     begin
24       if rising_edge(clock) then
25         if (ps2_clock = 1 ) then
26           if high=8 then
27             clk_internal <= 1 ;
28             high <= 0;
29             low <= 0
30           else
31             high <= high +1;
32           end if;
33         else
34           if low=8 then
35             clk_internal <= 0 ;
36             low <= 0;
37             high <= 0;
```

```
38          else
39              low <= low +1;
40          end if;
41        end if;
42      end if;
43    end process;
44    process(clk_internal) is
45    begin
46      if falling_edge(clk_internal) then
47        d(index) <= data;
48        if index>0 then
49          index <= index-1;
50        else
51          byte1 <= d(23 downto 16);
52          byte2 <= d(15 downto 8);
53          byte3 <= d(7 downto 0);
54          index<=23;
55        end if;
56      end if;
57    end process;
58  end architecture basic;
```

In this case the modified mouse handler waits for eight consecutive highs or lows on the clock signal at the higher internal clock rate of the FPGA and then it will set the internal clock high or low, respectively. Then the same mouse handler routine takes over to manage the data input, this time using the internally generated clock.

12.9 Basic PS/2 Mouse Handler in Verilog

The simplest form of the Verilog handler could use the keyboard clock signal as the system clock and then monitor the data coming from the keyboard. This is shown here:

```
1   module psmouse (
2   clk,  // clock input
3   data  // data input
4   );
5
6   input clk;
7   input data;
8
9   wire clk;
10  wire data;
11
12  reg [4:0] index = 5'b10111;
13  reg [23:0] d;
14  reg [7:0] byte1;
15  reg [7:0] byte2;
16  reg [7:0] byte3;
17
18  always @ (negedge clk)
19  begin : count
20    d[index] <= data;
21    if (index > 0) begin
```

```
22        index <= index - 1;
23      end
24      else begin
25        index <= 23;
26        byte1 <= d[23:16];
27        byte2 <= d[15:8];
28        byte3 <= d[7:0]
29      end
30    end
31
32    endmodule
```

This Verilog is very simple: on each falling edge of the clock the current value of the data is read into the next element of the data array (d) and when the complete 24-bits packet has been read in (and index has counted down to zero) then the three bytes are transcribed from the packet.

A modified handler could also be implemented in Verilog, either using the same approach as described for the VHDL model, but also the approach could be taken to simply divide down the clock from a higher frequency reference.

12.10 Summary

This chapter has shown how to handle a basic PS/2 signal for a mouse and then store the data in three bytes for further processing. Two methods are shown for collecting the data, one using the PS/2 clock and the other using a sampled version with a much faster internal clock.

CHAPTER 13

PS/2 Keyboard Interface

13.1 Introduction

The PS/2 keyboard is a standard interface to both computers and also many FPGA development kits. The protocol is a serial one. In this chapter the basics of the protocol will be reviewed and also a simple VHDL interface code to enable the designer to use a PS/2 keyboard, primarily on a standard FPGA development kit.

13.2 PS/2 Keyboard Basics

The origins of the PS/2 keyboard are back in the 1980s with the proliferation of the IBM Personal Computer (PC). This had the generic name of Personal System; hence, PS and the second version of this was therefore called the PS/2. The interface technology has remained under that name ever since. The keyboard interface evolved from the XT (83 key, 5pin DIN), through the AT (84-101 key, 5pin DIN) and eventually settled on the PS/2 (84-101 key, 6pin miniDIN).

The PS/2 interface is essentially a custom serial interface with one device supported per connector (unlike the modern USB, Universal Serial Bus, which can handle numerous devices on a single port). The data rate is relatively slow (40 kbps) and the device is powered off a 5 V dc supply.

Unlike the USB approach where devices are generally "hot swappable," that is, they can be plugged in or unplugged at will, the PS/2 device cannot be removed without a system crash or freeze resulting.

The PS/2 keyboard supports communication from the keyboard to the host and vice versa, and the supply is provided from the host to the keyboard in the form of a 5 V line.

Unlike the mouse, the keyboard has an on-board processor that checks the matrix of keys for any key presses and sends the appropriate code down the PS/2 data line.

13.3 PS/2 Keyboard Commands

The keyboard processor has two commands that are sent to the host system when a key is pressed: the make and the break command. Each key has a separate code that is sent in each case. The code that is actually sent to the host has no relationship to the ASCII code of the character sent. It is up to the host code to decode the key command sent. For example, the character 5 has the make code 0x2E and the break code 0xF0,0x2E. Most standard characters have a one-byte make code and a two-byte break code, and extended characters often have two-byte make codes and three-byte break codes.

If a key is pressed, then the make code is sent periodically until another key is pressed. The rate of this is called the typematic rate and is defined as default at approximately 10 characters per second.

13.4 PS/2 Keyboard Data Packets

The PS/2 keyboard sends data in serial packets down a data line and this is synchronous with a clock line also on the mouse interface. Each packet consists of up to three, 8-bit bytes and this can be decoded by a look-up table (LUT) for the keyboard scan codes.

13.5 PS/2 Keyboard Operation Modes

13.5.1 Basic PS/2 Keyboard Handler in VHDL

The simplest form of the VHDL handler could use the keyboard clock signal as the system clock and then monitor the data coming from the keyboard. This is shown here:

```
1    library ieee;
2    use ieee.std_logic_1164.all;
3
4    entity pskeyboard is
5      port (
6        clock : in std_logic;
7        data : in std_logic
8        );
9    end entity pskeyboard;
10
11   architecture basic of pskeyboard is
12     signal d : std_logic_vector (23 downto 0);
13     signal byte1 : std_logic_vector (7 downto 0);
14     signal byte2 : std_logic_vector (7 downto 0);
15     signal byte3 : std_logic_vector (7 downto 0);
16     signal index : integer := 23;
```

```
17    begin
18      process(clock) is
19      begin
20        if falling_edge(clock) then
21          d(index) <= data;
22          if index>0 then
23            index <= index−1;
24          else
25            byte1 <= d(23 downto 16);
26            byte2 <= d(15 downto 8);
27            byte3 <= d(7 downto 0);
28            index<=23;
29          end if;
30        end if;
31      end process;
32    end architecture basic;
```

This VHDL is very simple. On each falling edge of the clock the current value of the data is read into the next element of the data array (d) and when the complete 24-bits packet has been read in (and index has counted down to zero) then the three bytes are transcribed from the packet.

13.5.2 Modified PS/2 Keyboard Handler in VHDL

The trouble with the previous keyboard handler is that, although syntactically correct, there could be noise on the keyboard clock and data signals leading to an incorrect clocking of the data and so another approach would be to have a much higher frequency signal clock and to monitor the PS/2 clock as if it were a signal. An extra check would be to filter the PS/2 clock so that only if there were a certain number of the same values would the clock be considered to have changed.

```
1     library ieee;
2     use ieee.std_logic_1164.all;
3
4     entity pskeyboard is
5      port (
6         clk : in std_logic;
7         ps2_clock : in std_logic;
8         data : in std_logic
9      );
10    end entity pskeyboard;
11
12    architecture basic of pskeyboard is
13      signal clk_internal : std_logic := 0 ;
14      signal d : std_logic_vector (23 downto 0);
15      signal byte1 : std_logic_vector (7 downto 0);
16      signal byte2 : std_logic_vector (7 downto 0);
17      signal byte3 : std_logic_vector (7 downto 0);
```

```
18          signal index : integer := 23;
19      begin
20        process(clock) is
21          high : integer := 0;
22          low : integer := 0;
23        begin
24          if rising_edge(clock) then
25            if (ps2_clock = 1 ) then
26              if high=8 then
27                clk_internal <= 1 ;
28                high <= 0;
29                low <= 0
30              else
31                high <= high +1;
32              end if;
33            else
34              if low=8 then
35                clk_internal <= 0 ;
36                low <= 0;
37                high <= 0;
38              else
39                low <= low +1;
40              end if;
41            end if;
42          end if;
43        end process;
44        process(clk_internal) is
45        begin
46          if falling_edge(clk_internal) then
47            d(index) <= data;
48            if index>0 then
49              index <= index-1;
50            else
51              byte1 <= d(23 downto 16);
52              byte2 <= d(15 downto 8);
53              byte3 <= d(7 downto 0);
54              index<=23;
55            end if;
56          end if;
57        end process;
58      end architecture basic;
```

In this case the modified keyboard handler waits for eight consecutive highs or lows on the clock signal at the higher internal clock rate of the FPGA and then it will set the internal clock high or low, respectively. Then the same keyboard handler routine takes over to manage the data input, this time using the internally generated clock.

13.5.3 Basic PS/2 Keyboard Handler in Verilog

The simplest form of the Verilog handler could use the keyboard clock signal as the system clock and then monitor the data coming from the keyboard; this is shown here:

```
1    module pskeyboard (
2      clk,          // clock input
3      data          // data input
4      );
5
6    input clk;
7    input data;
8
9    wire clk;
10   wire data;
11
12   reg [4:0] index = 5'b10111;
13   reg [23:0] d;
14   reg [7:0] byte1;
15   reg [7:0] byte2;
16   reg [7:0] byte3;
17
18   always @ (negedge clk)
19   begin : count
20     d[index] <= data;
21     if (index > 0) begin
22       index <= index - 1;
23     end
24     else begin
25       index <= 23;
26       byte1 <= d[23:16];
27       byte2 <= d[15:8];
28       byte3 <= d[7:0];
29     end
30   end
31
32   endmodule
```

This Verilog is very simple: On each falling edge of the clock the current value of the data is read into the next element of the data array (d) and when the complete 24-bits packet has been read in (and index has counted down to zero) then the three bytes are transcribed from the packet.

A modified handler could also be implemented in Verilog, either using the same approach as described for the VHDL model, but also the approach could be taken to simply divide down the clock from a higher frequency reference.

13.6 Summary

This chapter has shown how to handle a basic PS/2 signal for a keyboard and then store the data in three bytes for further processing. Two methods are shown for collecting the data, one using the PS/2 clock and the other using a sampled version with a much faster internal clock.

CHAPTER 14

A Simple VGA Interface

14.1 Introduction

The VGA interface is common to most modern computer displays and is based on a pixel map, color planes, and horizontal and vertical sync signals. A VGA monitor has three color signals (Red, Green, and Blue) that set one of these colors on or off on the screen. The intensity of each of those colors sets the final color seen on the display. For example, if the Red was fully on, but the Blue and Green off, then the color would be seen as a strong red. Each analog intensity is defined by a 2-bit digital word for each color (e.g., red0 and red1) that are connected to a simple digital-to-analog converter to obtain the correct output signal.

The resolution of the screen can vary from 480×320 up to much larger screens, but a standard default size is 640×480 pixels. This is 480 lines of 640 pixels in each line, so the aspect ratio is 640/480, leading to the classic landscape layout of a conventional monitor screen.

The VGA image is controlled by two signals—horizontal sync and vertical sync. The horizontal sync marks the start and finish of a line of pixels with a negative pulse in each case. The actual image data is sent in a 25.17 µs window in a 31.77 µs space between the sync pulses. (The time that image data is not sent is where the image is defined as a blank space and the image is dark.) The vertical sync is similar to the horizontal sync except that in this case the negative pulse marks the start and finish of each frame as a whole and the time for the frame (image as a whole) takes place in a 15.25 ms window in the space between pulses, which is 16.784 ms.

There are some constraints about the spacing of the data between pulses which will be considered later in this chapter, but it is clear that the key to a correct VGA output is the accurate definition of timing and data by the VHDL.

14.2 Basic Pixel Timing

If there is a space of 25.17 s to handle all of the required pixels, then some basic calculations need to be carried out to make sure that the FPGA can display the correct data in the time available. For example, if we have a 640×480 VGA system, then that means that 640 pixels must be sent to the monitor in 25.17 µs. Doing the simple calculation shows that for each pixel

Design Recipes for FPGAs. http://dx.doi.org/10.1016/B978-0-08-097129-2.00014-3

197

we need 25.17 µs/640 = 39.328 ns per pixel. If our clock frequency is 100 MHz on the FPGA, then that gives a minimum clock period of 10 ns, so this can be achieved with a relatively standard FPGA.

14.3 Image Handling

Clearly it is not sensible to use an integrated image system on the FPGA, but rather it makes much more sense to store the image in memory (RAM) and retrieve it frame by frame. Therefore, as well as the basic VGA interface, it makes a lot of sense for the images to be moved around in memory and using the same basic RAM interface as defined previously is sensible. Therefore, as well as the VGA interface pins, our VGA handler should include a RAM interface.

14.4 A VGA Interface in VHDL

14.4.1 VHDL Top Level Entity for VGA Handling

The first stage in defining the VHDL for the VGA driver is to create a VHDL entity that has the global clock and reset, the VGA output pins, and a memory interface. The outline VHDL entity is therefore given as follows:

```
1     library ieee;
2     use ieee.std_logic_1164.all;
3     entity vga is
4       port (
5         clk : in std_logic;
6         nrst : in std_logic;
7         hsync : out std_logic;
8         vsync : out std_logic;
9         red : out std_logic_vector ( 1 downto 0);
10        green : out std_logic_vector ( 1 downto 0);
11        blue : out std_logic_vector ( 1 downto 0);
12        address : out (std_logic_vector ( 15 downto 0);
13        data : in (std_logic_vector ( 7 downto 0);
14        ram_en : out std_logic;
15        ram_oe : out std_logic;
16        ram_wr : out std_logic
17      );
18    end entity vga;
19
20    architecture core of vga is
21
22      -- vga internal signals go here
23
24    begin
25
26      -- vga interface core goes here
27
28    end architecture core;
```

The architecture contains a number of processes, with internal signals that manage the transfer of pixel data from memory to the screen. As can be seen from the entity, the data comes back from the memory in 8-bit blocks and we require 3×2 bits for each pixel and so when the data is returned, each memory byte will contain the data for a single pixel. In this example, as we are using a 640×480 pixel image, this will therefore require a memory that is 307,200 bytes in size as a minimum. To put this in perspective, this means that using a raw memory approach we can put three frames per megabyte. In practice, of course, we would use a form of image compression (such as JPEG for photographic images), but this is beyond the scope of this book.

We can therefore use a simple process to obtain the current pixel of data from memory as follows:

```
1     mem_read : process ( pclk, nrst ) is
2       signal current_address : unsigned (16 downto 0);
3     begin
4       if nrst =    0      then
5         pixelcount <= 0;
6         current_address <= 0;
7       else
8         if rising_edge(pclk) then
9           current_address <= current_address + 1;
10          address <= std_logic_vector(current_address);
11          pixel_data <= data;
12        end if;
13      end if;
14    end process;
```

This process returns the current value of the pixel data into a signal called pixel_data, which is declared at the architecture level:

```
1     signal pixel_data : std_logic_vector ( 7 downto 0 );
```

This has the red, green, and blue data defined in the lowest 6 bits of the 8-bit data word with the indexes, respectively, of 0-1, 2-3, and 4-5.

14.4.2 Horizontal Sync

The next key process is the timing of the horizontal and vertical sync pulses, and the blanking intervals. The line timing for VGA is 31,770 ns per line with a window for displaying the data of 25,170 ns. If the FPGA is running at 100 MHz (period of 10 ns) then this means that each line requires 3177 clock cycles with 2517 for each line of pixel data, with 660 pulses in total for blanking (330 at either side). This also means that for a 640 pixel wide line, 39.3 ns are required for each pixel. We could round this up to 4 clock cycles per pixel. As you may have noticed, for the pixel retrieval we have a new internal clock signal called pclk, and we can create a process that generates the appropriate pixel clock (pclk) with this timing in place.

With this slightly elongated window, the blanking pulses must therefore reduce to 617 clock cycles and this means 308 before and 309 after the display window.

The horizontal sync pulse, on the other hand, takes place between 26,110 ns and 29,880 ns of the overall interval. This is 189 clock pulses less than the overall line time, and so the horizontal sync pulse goes low after 94 clock cycles and then at the end must return high 95 clock cycles prior to the end of the line. The difference between the outside and inside timings for the horizontal sync pulse is 377 clock cycles and so the sync pulse must return high $94 + 188$ clock cycles and then return low $95 + 189$ prior to the end of the window.

Thus, the horizontal sync has the following basic behavior:

Clock	Cycle Value
0	1
94	0
282	1
2893	0
3082	1

and this can be implemented using a process with a simple counter:

```
1    hsync_counter : process (clk, nrst ) is
2      hcount : unsigned ( 11 downto 0 );
3    begin
4      if nrst =    0       then
5        hcount <= 0;
6        hsync <=    1;
7      else
8        if hcount > 2611 and hcount<2988 then
9          hsync <=    0;
10       else
11         hsync <=    1;
12       end if;
13       if hcount < 3177 then
14         hcount <= hcount + 1;
15       else
16         hcount <= 0;
17       end if;
18     end if;
19   end process;
```

14.4.3 Vertical Sync

The horizontal sync process manages the individual pixels in a line, and the vertical sync does the same for the lines as a whole to create the image. The period of a frame (containing all the lines) is defined as 16,784,000 ns. Within this timescale, the lines of the image are displayed (within 15,250,000 ns), then the vertical blanking interval is defined (up to the whole frame period of 16,784,000 ns) and finally the vertical sync pulse is defined as 1 until 15,700,000 ns at which time it goes to zero, returning to 1 at 15,764,000 ns.

Clearly it would not be sensible to define a clock of 10 ns for these calculations, so the largest common divisor is a clock of 2 s, so we can divide down the system clock by 2000 to get a vertical sync clock of 2 s to simplify and make the design as compact as possible.

```
1    clk_div : process (clk, nrst ) is
2    begin
3     if nrst =   0       then
4       count <= 0;
5       vclk <=    0;
6     else
7       if count = 1999 then
8         count <= 0;
9         vclk <= not vclk;
10      else
11        count <= count + 1;
12      end if;
13    end if;
14   end process;
```

where the vertical sync clock (vclk) is defined as a std_logic signal in the architecture. This can then be used to control the vsync pulses in a second process that now waits for the vertical sync derived clock:

```
1    vsync_timing : process (vclk) is
2    begin
3     if nrst =   0       then
4       vcount <= 0;
5     else
6       if vcount>15700 and vcount <15764 then
7         vsync <=    0;
8       else
9         vsync <=    1;
10      end if;
11      if vcount > 16784 then
12        vcount <= 0;
13      else
14        vcount <= vcount + 1;
15      end if;
16    end if;
17   end process;
```

Using this process, the vertical sync (frame synchronization) pulses are generated.

14.4.4 Horizontal and Vertical Blanking Pulses

In addition to the basic horizontal and vertical sync pulse counters, we have to define a horizontal blanking pulse which sets the line data low after 25,170 ns (2517 clock cycles). This can be implemented as a simple counter in exactly the same way as the horizontal sync

pulse and similarly for a vertical blanking pulse. The two processes to implement these are
given in the following VHDL.

```
1     hblank_counter : process (clk, nrst ) is
2       hcount : unsigned ( 11 downto 0 );
3     begin
4       if nrst =    0       then
5         hcount <= 0;
6         hblank <=    1;
7       else
8         if hcount > 2517 and hcount<3177 then
9           hblank <=    0;
10        else
11          hblank <=    1;
12        end if;
13        if hcount < 3177 then
14          hcount <= hcount + 1;
15        else
16          hcount <= 0;
17        end if;
18      end if;
19    end process;
20    vblank_timing : process (vclk) is
21    begin
22      if nrst =    0       then
23        vcount <= 0;
24        vblank<=    1      1
25      else
26        if vcount>15250 and vcount <16784 then
27          vblank <=    0;
28        else
29          vblank <=    1;
30        end if;
31        if vcount > 16784 then
32          vblank <= 0;
33        else
34          vcount <= vcount + 1;
35        end if;
36      end if;
37    end process;
```

14.4.5 Calculating the Correct Pixel Data

As we have seen previously, the data of each pixel is retrieved from a memory location and
this is obtained using the pixel clock (pclk). The pixel clock is simply a divided (by 4) version
of the system clock and at each rising edge of this pclk signal, the next pixel data is obtained
from the memory data stored in the signal called data and translated into the red, green, and
blue line signals. This is handled using the basic process given here:

```
1     pixel_handler : process (pclk) is
2     begin
3       red <= data(1 downto 0);
4       green <= data(3 downto 2);
```

```
5      blue <= data(5 downto 4);
6    end process;
```

This is a basic handler process that picks out the correct pixel data, but it does not include the video blanking signal and if this is included, then the simple VHDL changes slightly to this form:

```
1    pixel_handler : process (pclk) is
2      blank : std_logic_vector (1 downto 0);
3    begin
4      blank(0) <= hblank or vblank;
5      blank(1) <= hblank or vblank;
6      red <= data(1 downto 0) & blank;
7      green <= data(3 downto 2) & blank;
8      blue <= data(5 downto 4) & blank;
9    end process;
```

This is the final step and completes the basic VHDL VGA handler.

14.5 A VGA Interface in Verilog

14.5.1 Verilog Top Level Module for VGA Handling

The first stage in defining the Verilog for the VGA driver is to create a module that has the global clock and reset, the VGA output pins, and a memory interface. The outline Verilog module is therefore given as follows:

```
1    module vga (
2      clk, // clock input
3      nrst, // reset
4      hsync, // hsync
5      vsync, // vsync
6      red, // red output
7      green, // green output
8      blue, // blue output
9      address, // address output
10     data, // data
11     ram_en, // RAM enable
12     ram_oe, // RAM Output Enable
13     ram_wr // RAM write signal
14     );
```

The module contains a number of processes, with internal signals that manage the transfer of pixel data from memory to the screen. As before, the data comes back from the memory in 8-bit blocks and we require 3×2 bits for each pixel and so when the data is returned, each memory byte will contain the data for a single pixel. In this example, as we are using a 640×480 pixel image, this will therefore require a memory that is 307,200 bytes in size as a minimum. To put this in perspective, this means that using a raw memory approach we can put three frames per megabyte. In practice, of course, we would use a form of image compression (such as JPEG for photographic images), but this is beyond the scope of this book.

We can therefore use a simple process to obtain the current pixel of data from memory as follows:

```
1    always @ (posedge pclk)
2    begin
3     if (nrst = 0) begin
4       pixelcount <= 0;
5       current_address <= 0;
6     else
7       current_address <= current_address + 1;
8       address <= current_address;
9       pixel_data <= data;
10
11    end
12   end
```

This process returns the current value of the pixel data into a signal called pixel_data which is declared at the module level:

```
1    reg [7:0] pixel_data;
```

This has the red, green, and blue data defined in the lowest 6 bits of the 8-bit data word with the indexes, respectively, of 0-1, 2-3, and 4-5.

14.5.2 Horizontal Sync

The next key process is the timing of the horizontal and vertical sync pulses, and the blanking intervals. The line timing for VGA is 31,770 ns per line with a window for displaying the data of 25,170 ns. If the FPGA is running at 100 MHz (period of 10 ns) then this means that each line requires 3177 clock cycles with 2517 for each line of pixel data, with 660 pulses in total for blanking (330 at either side). This also means that for a 640 pixel wide line, 39.3 ns are required for each pixel. We could round this up to 4 clock cycles per pixel. As you may have noticed, for the pixel retrieval we have a new internal clock signal called pclk, and we can create a process that generates the appropriate pixel clock (pclk) with this timing in place.

With this slightly elongated window, the blanking pulses must therefore reduce to 617 clock cycles and this means 308 before and 309 after the display window.

The horizontal sync pulse, on the other hand, takes place between 26,110 ns and 29,880 ns of the overall interval. This is 189 clock pulses less than the overall line time, and so the horizontal sync pulse goes low after 94 clock cycles and then at the end must return high 95 clock cycles prior to the end of the line. The difference between the outside and inside timings for the horizontal sync pulse is 377 clock cycles and so the sync pulse must return high 94 + 188 clock cycles and then return low 95 + 189 prior to the end of the window.

Thus, the horizontal sync has the same behavior as described previously in this chapter and this can be implemented using a process with a simple counter:

```
1    always @ (posedge clk)
2    begin
3      if (nrst = 0) begin
4        hcount <= 0;
5        hsync <= 1;
6      end
7      else
8        if (hcount > 2611) and (hcount<2988) begin
9          hsync <= 0;
10       else
11         hsync <= 1;
12       end if
13       if (hcount < 3177) begin
14         hcount <= hcount + 1;
15       else
16         hcount <= 0;
17       end if
18     end if
19   end
```

14.5.3 Vertical Sync

The horizontal sync process manages the individual pixels in a line, and the vertical sync does the same for the lines as a whole to create the image. The period of a frame (containing all the lines) is defined as 16,784,000 ns. Within this timescale, the lines of the image are displayed (within 15,250,000 ns), then the vertical blanking interval is defined (up to the whole frame period of 16,784,000 ns) and finally the vertical sync pulse is defined as 1 until 15,700,000 ns at which time it goes to zero, returning to 1 at 15,764,000 ns.

Clearly it would not be sensible to define a clock of 10 ns for these calculations, so the largest common divisor is a clock of 2 s, so we can divide down the system clock by 2000 to get a vertical sync clock of 2 s to simplify and make the design as compact as possible.

```
1    always @ (posedge clk) begin
2      if (nrst = 0) begin
3        count <= 0;
4        vclk <= 0;
5      else
6        if (count = 1999) begin
7          count <= 0;
8          vclk <= not vclk;
9        else
10         count <= count + 1;
11       end if
12     end if
13   end
```

where the vertical sync clock (vclk) is defined as a std_logic signal in the architecture. This can then be used to control the vsync pulses in a second process that now waits for the vertical sync derived clock:

```
1      always @ (vclk) begin
2        if nrst =    0       begin
3          vcount <= 0;
4        else
5          if vcount>15700 and vcount <15764 begin
6            vsync <=    0;
7          else
8            vsync <=    1;
9          end if
10         if vcount > 16784 begin
11           vcount <= 0;
12         else
13           vcount <= vcount + 1;
14         end if
15       end if
16     end
```

Using this process, the vertical sync (frame synchronization) pulses are generated.

14.5.4 Horizontal and Vertical Blanking Pulses

In addition to the basic horizontal and vertical sync pulse counters, we have to define a horizontal blanking pulse, which sets the line data low after 25,170 ns (2517 clock cycles). This can be implemented as a simple counter in exactly the same way as the horizontal sync pulse and similarly for a vertical blanking pulse. The two processes to implement these are given in the following VHDL.

```
1      always @ (posedge clk) begin
2
3        if nrst =    0       begin
4          hcount <= 0;
5          hblank <=    1;
6        end
7        else
8          if hcount > 2517 and hcount<3177 begin
9            hblank <=    0;
10         else
11           hblank <=    1;
12         end if
13         if hcount < 3177 then
14           hcount <= hcount + 1;
15         else
16           hcount <= 0;
17         end if
18       end if
19     end
20
21     always @ (posedge vclk) begin
22       if nrst =    0       begin
23         vcount <= 0;
24         vblank<=    1      1
25       end
26       else
```

```
27        if vcount>15250 and vcount <16784 begin
28          vblank <=    0;
29        else
30          vblank <=    1;
31        end if
32        if vcount > 16784 begin
33          vblank <= 0;
34        else
35          vcount <= vcount + 1;
36        end if
37      end
38    end
```

14.5.5 Calculating the Correct Pixel Data

As we have seen previously, the data of each pixel is retrieved from a memory location and this is obtained using the pixel clock (pclk). The pixel clock is simply a divided (by 4) version of the system clock and at each rising edge of this pclk signal, the next pixel data is obtained from the memory data stored in the signal called data and translated into the red, green and blue line signals. This is handled using a similar process as was used for the VHDL.

This is the final step and completes the basic Verilog VGA handler.

14.6 Summary

This chapter has introduced the basics of developing a simple VGA handler in VHDL and Verilog. While it is a simplistic view of the process, hopefully it has shown how a simple VGA interface can be developed using not very complex VHDL or Verilog and a building block approach. It is left to the reader to develop their own complete VGA routines for the specific monitor that they have, using the techniques developed in this chapter as a basis.

14.5.5 Calculating the Correct Pixel Data

14.6 Summary

CHAPTER 15

Serial Communications

15.1 Introduction

There are a wide variety of serial communications protocols available, but all rely on some form of coding scheme to efficiently and effectively transmit the serial data across the transmission medium. In this chapter, not only will the common methods of transmitting data be reviewed (RS232 and USB), but in addition some useful coding mechanisms will be described (Manchester, Code Mark Inversion, Non-Return-to-Zero—NRZ, Non-Return-to-Zero-Inverted—NRZI) as they often are used as part of a higher level transmission protocol. For example, the NRZI coding technique is used in the USB protocol.

15.2 Manchester Encoding and Decoding

Manchester encoding is a simple coding scheme that translates a basic bit stream into a series of transitions. It is extremely useful for ensuring that a specific bandwidth can be used for data transmission, as no matter what the sequence of the data bits, the frequency of the transmitted stream will be exactly twice the frequency of the original data. It also makes signal recovery trivial, because there is no need to attempt to extract a clock as the data can be recovered simply by looking for the edges in the data and extracting asynchronously. The basic approach to Manchester encoding is shown in Figure 15.1.

Another advantage of the scheme is that the method is highly tolerant of errors in the data; if an error occurs, then the subsequent data is not affected at all by an error in the transmitter, the medium or the receiver, and after the immediate glitch, the data can continue to be transmitted effectively without any need for error recovery. Of course, the original data can use some form of data coding to add in error correction (such as parity checks or cyclic redundancy check, CRC). If we wish to create a VHDL model for this type of coding scheme, it is actually relatively simple. The first step is to identify that we have a single data input (D) and a clock (CLK). Why synchronous? Using a synchronous clock we can define a sample on the rising edge of the clock for the data input and use BOTH edges of the clock to define the transitions on the output. The way we do this is simply to look for any event on the clk (clk'event) and then check whether the clk is high or low to determine whether the clk edge is rising or falling.

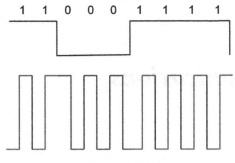

Figure 15.1
Manchester encoding scheme.

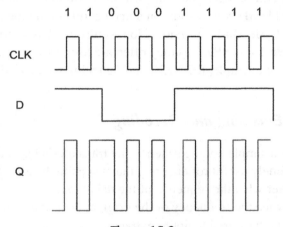

Figure 15.2
Manchester encoding scheme with XOR.

This VHDL is simple but there is an even simpler way to encode the data and that is to simply XOR the clock with the data. If we look at the same data sequence as shown in Figure 15.1, we can see that if we add a clock, and observe the original data and the Manchester encoded output, this is simply the data XORd with the clock, as in Figure 15.2.

So, using this simple mechanism, we can create a much simpler Manchester encoder that XORs the clock and the data to obtain the resulting Manchester encoded data stream.

15.3 Implementing the Manchester Encoding Scheme using VHDL

The VHDL code for the synchronous Manchester encoder is shown here:

```
1    library ieee;
2    use ieee.std_logic_1664.all;
3
4    entity manchester_encoder is
5      port (
6        clk : in std_logic;
7        d : in std_logic;
8        q : out std_logic
9        );
10   end entity manchester_encoder;
11
12   architecture basic of manchester_encoder is
13     signal lastd : std_logic := 0 ;
14   begin
15     p1: process ( clk )
16     begin
17       if clk'event and clk='1' then
18         if ( d = 0 ) then
19           q <= 1 ;
20           lastd <= 0 ;
21         elsif ( d = 1 ) then
22           q <= 0 ;
23           lastd <= 1 ;
24         else
25           q <= x ;
26           lastd <= x ;
27         end if;
28       else
29         if ( lastd = 0 ) then
30           q <= 0 ;
31         elsif ( lastd = 1 ) then
32           q <= 1 ;
33         else
34           q <= x ;
35         end if
36       end if;
37     end process p1;
38   end architecture basic;
```

The XOR implementation of the Manchester encoder is much simpler and this listing is shown as follows:

```
1    library ieee;
2    use ieee.std_logic_1664.all;
3
4    entity manchester_encoder is
5      port (
6        clk : in std_logic;
7        d : in std_logic;
8        q : out std_logic
9        );
10   end entity manchester_encoder;
11
```

```
12    architecture basic of manchester_encoder is
13    begin
14      q <= d xor clk;
15    end architecture basic;
```

Decoding the Manchester data stream is also a choice between asynchronous and synchronous approaches. We can use a local clk and detect the values of the input to evaluate whether the values on the rising and falling edges are 0 or 1, respectively, and ascertain the values of the data as a result, but clearly this is dependent on the transmitter and receiver clocks being synchronized to a reasonable degree. Such a simple decoder could look like this:

```
1     entity manchester_decoder is
2       port (
3         clk : in std_logic;
4         d : in std_logic;
5         q : out std_logic
6       );
7     end entity manchester_decoder;
8
9     architecture basic of manchester_decoder is
10      signal lastd : std_logic := 0 ;
11    begin
12      p1 : process (clk)
13      begin
14        if clk'event and clk='1' then
15          lastd <= d;
16        else
17          if (lastd = 0 ) and (d = 1 ) then
18            q <= 1 ;
19          elsif (lastd = 1 ) and (d= 0 ) then
20            q <= 0 ;
21          else
22            q <= x ;
23          end if;
24        end if;
25      end process p1;
26    end architecture basic;
```

In this VHDL model, the clock is at the same rate as the transmitter clock, and the data should be sent in packets to ensure that the data is not sent in blocks that are too large, such that the clock can get out of sync, and also that the data can be checked for integrity to correct for mistakes or the clock on the receiver being out of phase.

15.4 Implementing the Manchester Encoding Scheme using Verilog

The Verilog code for the synchronous Manchester encoder is shown here:

```
1     module manchester (
2       clk, // clock input
3       d, // data input
4       q  // encoded output
```

```
5    );
6
7    input clk;
8    input d;
9
10   output q;
11   reg q;
12   reg lastd;
13
14   always_init begin
15     lastd <= 0;
16   end
17
18   always @ (clk)
19   begin
20     if clk='1' begin
21       if d=1 then
22       begin
23         q <= 1;
24         lastd <= 0;
25       end
26       else
27       begin
28         q <= 0;
29         lastd <= 1;
30       end
31     end
32     else
33     begin
34       if lastd = 0 then
35         q <= 0;
36       else
37         q = 1;
38     end
39   end
40   endmodule
```

The XOR implementation of the Manchester encoder is much simpler and this listing is shown as follows:

```
1    module manchester_xor (
2      clk, // clock input
3      d, // data input
4      q // encoded output
5    );
6
7    input clk;
8    input d;
9
10   output q;
11   reg q;
12
13   q <= clk xor d;
14
15   endmodule
```

Decoding the Manchester data stream is also a choice between asynchronous and synchronous approaches. We can use a local clk and detect the values of the input to evaluate whether the values on the rising and falling edges are 0 or 1, respectively, and ascertain the values of the data as a result, but clearly this is dependent on the transmitter and receiver clocks being synchronized to a reasonable degree. Such a simple decoder could look like this:

```
1    entity manchester_decoder is
2      port (
3        clk : in std_logic;
4        d : in std_logic;
5        q : out std_logic
6      );
7    end entity manchester_decoder;
8
9    architecture basic of manchester_decoder is
10     signal lastd : std_logic := 0 ;
11   begin
12     p1 : process (clk)
13     begin
14       if clk'event and clk='1' then
15         lastd <= d;
16       else
17         if (lastd = 0 ) and (d = 1 ) then
18           q <= 1 ;
19         elsif (lastd = 1 ) and (d= 0 ) then
20           q <= 0 ;
21         else
22           q <= x ;
23         end if;
24       end if;
25     end process p1;
26   end architecture basic;
```

In this Verilog model, the clock is at the same rate as the transmitter clock, and the data should be sent in packets to ensure that the data is not sent in too large blocks such that the clock can get out of sync, and also that the data can be checked for integrity to correct for mistakes or the clock on the receiver being out of phase.

15.5 NRZ (Non-Return-to-Zero) Coding and Decoding

The NRZ encoding scheme is actually not a coding scheme at all. It simply states that a 0 is transmitted as a 0 and a 1 is transmitted as a 1. It is only worth mentioning because a designer may see the term NRZ and assume that a specific encoder or decoder was required, whereas in fact this is not the case. It is also worth noting that there are some significant disadvantages in using this simple approach. The first disadvantage, especially when compared to the Manchester coding scheme, is that long sequences of 0s or 1s give effectively DC values when transmitted, which are susceptible to problems of noise and also make clock recovery very difficult. The other issue is that of bandwidth. Again if we compare the coding scheme to that of the Manchester example, it is obvious that the Manchester scheme requires quite a narrow bandwidth (relatively) to transmit the data, whereas the NRZ scheme may require anything from DC up to half the data rate (Nyquist frequency) and anything in between. This makes line design and filter design very much more problematic.

15.6 NRZI (Non-Return-to-Zero-Inverted) Coding and Decoding

In the NRZI scheme, the potential problems of the NRZ scheme, particularly the long periods of DC levels, are partially alleviated. In the NRZI, if the data is a 0, then the data does not change, whereas if a 1 occurs on the data line, then the output changes. Therefore, the issue of long sequences of 1s is addressed, but the potential for long sequences of 0s remains.

15.6.1 NRZI Coding and Decoding in VHDL

It is a simple matter to create a basic model for a NRZI encoder using the following VHDL model:

```
1    entity nrzi_encoder is
2      port (
3        clk : in std_logic;
4        d : in std_logic;
5        q : out std_logic
6      );
7    end entity nrzi_encoder;
8
9    architecture basic of nrzi_encoder is
10     signal qint : std_logic := 0 ;
11   begin
12     p1 : process (clk)
13     begin
14       if (d = 1 ) then
15         if ( qint = 0 ) then
16           qint <= 1 ;
17         else
18           qint <= 0 ;
19         end if;
20       end if;
21     end process p1;
22     q <= qint;
23   end architecture basic;
```

Notice that this model is synchronous, but if we wished to make it asynchronous, the only changes would be to remove the clk port and change the process sensitivity list from clk to d. We can apply the same logic to the output, to obtain the decoded data stream, using the VHDL as follows. Again we are using a synchronous approach:

```
1    entity nrzi_decoder is
2      port (
3        clk : in std_logic;
4        d : in std_logic;
5        q : out std_logic
6      );
7    end entity nrzi_decoder;
8
9    architecture basic of nrzi_decoder is
10     signal lastd : std_logic := 0 ;
11   begin
12     p1 : process (clk)
13     begin
14       if rising_edge(clk) then
15       if (d = lastd) then
16         q <= 0 ;
17       else
18         q <= 1 ;
19       end if;
20       lastd <= d;
21       end if;
22     end process p1;
23   end architecture basic;
```

The NRZI decoder is extremely simple, in that the only thing we need to check is whether the data stream has changed since the last clock edge. If the data has changed since the last clock, then we know that the data is a 1, but if the data is unchanged, then we know that it is a 0. Clearly we could use an asynchronous approach, but this would rely on the data checking algorithm downstream being synchronized correctly.

15.6.2 NRZI Coding and Decoding in Verilog

It is a simple matter to create a basic model for a NRZI encoder using the following Verilog model:

```
1    module nrzi_encoder (
2      clk, // CLock Input
3      d,   // Data Input
4      q    // Data Output
5      );
6
7    input clk;
8    input d;
9    output q;
10
11   reg q;
12
13   reg qint;
14
15   always_init
```

```
16    begin
17      qint <= 0;
18    end
19
20    always @ (clk)
21    begin
22      if d=1
23        if qint = 0
24          qint <= 1
25        else
26          qint <= 0
27        end
28
29    end
30    q <= qint;
31    end
32    endmodule
```

Notice that this model is synchronous, but if we wished to make it asynchronous, the only changes would be to remove the clk port and change the process sensitivity list from clk to d. We can apply the same logic to the output, to obtain the decoded data stream, using the Verilog that follows. Again we are using a synchronous approach:

```
1    module nrzi_encoder (
2    clk, // CLock Input
3    d,   // Data Input
4    q    // Data Output
5    );
6
7    input clk;
8    input d;
9    output q;
10
11   reg q;
12
13   always @ (clk)
14   begin
15     if clk=1 begin
16       if d=lastd
17         q <= 0;
18       else
19         q <= 1;
20       end
21     end
22   end
23   endmodule
```

The NRZI decoder is extremely simple, in that the only thing we need to check is whether the data stream has changed since the last clock edge. If the data has changed since the last clock, then we know that the data is a 1, but if the data is unchanged, then we know that it is a 0. Clearly we could use an asynchronous approach, but this would rely on the data checking algorithm downstream being synchronized correctly.

15.7 RS-232

15.7.1 Introduction

The basic approach of RS-232 serial transmission is that of a UART. UART stands for Universal Asynchronous Receiver/Transmitter. It is the standard method of translating a serial communication stream into the parallel form used by computers. RS-232 is a UART that has a specific standard defined for start, stop, break, data, parity, and pin names.

15.7.2 RS-232 Baud Rate Generator

The RS-232 is an asynchronous transmission scheme and so the correct clock rate must be defined prior to transmission to ensure that the data is transmitted and received correctly. The RS-232 baud rate can range from 1200 baud up to 115200 baud. This is based on a standard clock frequency of 14.7456 MHz, and this is then divided down by 8, 16, 28, 48, 96, 192, 384, and 768 to get the correct baud rates. We therefore need to define a clock divider circuit that can output the correct baud rate configured by a control word. We have obviously got 8 different ratios, and so we can use a 3-bit control word (baud[2:0]) plus a clock and reset to create the correct frequencies, assuming that the basic clock frequency is 14.7456 MHz (Figure 15.3).

The VHDL for this controller is given as follows and uses a single process to select the correct baud rate and another to divide down the input clock accordingly:

```
1    library ieee;
2    use ieee.std_logic_1164.all;
3    use ieee.std_logic_unsigned.all;
4
5    entity baudcontroller is
6        port(
```

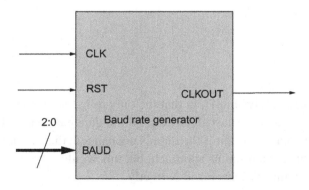

Figure 15.3
Baud rate generator.

```
 7          clk : in std_logic;
 8          rst : in std_logic;
 9          baud : in std_logic_vector(0 to 2);
10             clkout : out std_logic;
11      end baudcontroller;
12
13      architecture simple of baudcontroller is
14         signal clkdiv : integer := 0;
15         signal count : integer := 0;
16      begin
17        div: process (rst, clk)
18        begin
19         if rst='0' then
20           clkdiv <= 0;
21           count <= 0;
22         elsif rising_edge(clk) then
23           case baud is
24             when "000" => clkdiv <= 7;   -- 115200
25             when "001" => clkdiv <= 15;  -- 57600
26             when "010" => clkdiv <= 23;  -- 38400
27             when "011" => clkdiv <= 47;  -- 19200
28             when "100" => clkdiv <= 95;  -- 9600
29             when "101" => clkdiv <= 191; -- 4800
30             when "110" => clkdiv <= 383; -- 2400
31             when "111" => clkdiv <= 767; -- 1200
32             when others => clkdiv <= 7;
33           end case;
34         end if;
35        end process;
36
37        clockdivision: process (clk, rst)
38        begin
39         if rst='0' then
40           clkdiv <= 0;
41           count <= 0;
42         elsif rising_edge(clk) then
43           count <= count + 1;
44           if (count > clkdiv) then
45             clkout <= not clkout;
46             count <= 0;
47           end if;
48         end if;
49        end process;
50      end simple;
```

The Verilog equivalent for this baud rate controller is given as follows and uses separate code to select the correct baud rate and another to divide down the input clock accordingly:

```
1      module baudcontroller (
2        clk, // clock
3        rst, // reset
4        clkout // baud rate output
5      );
6
7      input clk;
8      input rst;
9      output clkout;
```

```
10     reg clkout;
11
12     parameter baudrate = 9600;
13
14     always_init
15       case baudrate
16         115200: clkdiv = 7;
17         57600: clkdiv = 15;
18         38400: clkdiv = 23;
19         19200: clkdiv = 47;
20         9600: clkdiv = 95;
21         4800: clkdiv = 191;
22         2400: clkdiv = 383;
23         1200: clkdiv = 767;
24
25
26             when "000" => clkdiv <= 7;   -- 115200
27             when "001" => clkdiv <= 15;  -- 57600
28             when "010" => clkdiv <= 23;  -- 38400
29             when "011" => clkdiv <= 47;  -- 19200
30             when "100" => clkdiv <= 95;  -- 9600
31             when "101" => clkdiv <= 191; -- 4800
32             when "110" => clkdiv <= 383; -- 2400
33             when "111" => clkdiv <= 767; -- 1200
34             when others => clkdiv <= 7;
35
36         always @(CharIn)
37         case (CharIn)
38           4'h0: HexOut = 7'b1000000;
39           4'h1: HexOut = 7'b1111001;
40           4'h2: HexOut = 7'b0100100;
41           4'h3: HexOut = 7'b0110000;
42           4'h4: HexOut = 7'b0011001;
43           4'h5: HexOut = 7'b0010010;
44           4'h6: HexOut = 7'b0000010;
45           4'h7: HexOut = 7'b1111000;
46           4'h8: HexOut = 7'b0000000;
47           4'h9: HexOut = 7'b0011000;
48           4'hA: HexOut = 7'b0001000;
49           4'hB: HexOut = 7'b0000011;
50           4'hC: HexOut = 7'b1000110;
51           4'hD: HexOut = 7'b0100001;
52           4'hE: HexOut = 7'b0000110;
53           4'hF: HexOut = 7'b0001110;
54           default: HexOut = 7'b0110110;
55         endcase
56
57     architecture simple of baudcontroller is
58         signal clkdiv : integer := 0;
59         signal count : integer := 0;
60     begin
61       div: process (rst, clk)
62       begin
63       if rst='0' then
64         clkdiv <= 0;
```

```
65           count <= 0;
66         elsif rising_edge(clk) then
67           case baud is
68             when "000" => clkdiv <= 7;   -- 115200
69             when "001" => clkdiv <= 15;  -- 57600
70             when "010" => clkdiv <= 23;  -- 38400
71             when "011" => clkdiv <= 47;  -- 19200
72             when "100" => clkdiv <= 95;  -- 9600
73             when "101" => clkdiv <= 191; -- 4800
74             when "110" => clkdiv <= 383; -- 2400
75             when "111" => clkdiv <= 767; -- 1200
76             when others => clkdiv <= 7;
77           end case;
78         end if;
79       end process;
80
81       clockdivision: process (clk, rst)
82       begin
83         if rst='0' then
84           clkdiv <= 0;
85           count <= 0;
86         elsif rising_edge(clk) then
87           count <= count + 1;
88           if (count > clkdiv) then
89             clkout <= not clkout;
90             count <= 0;
91           end if;
92         end if;
93       end process;
94     end simple;
```

15.7.3 RS-232 Receiver

The RS-232 receiver must wait for data to arrive on the RX line and has a specification defined as follows: <number of bits><parity><stop bits>. So, for example an 8-bit, No parity, 1 stop bit specification would be given as 8N1. The RS-232 voltage levels are between −12V and +12V, and so we will assume that an interface chip has translated these to standard logic levels (0-5 V or 0-3.3 V, for example). A sample bit stream would be of the format shown in Figure 15.4.

The idle state for RS-232 is high, and in this figure, after the stop bit, the line is shown as going low, when in fact that only happens when another data word is coming. If the data transmission has finished, then the line will go high (idle) again. We can in fact model this as a simple state machine as shown in Figure 15.5.

Figure 15.4
Example serial data bit stream.

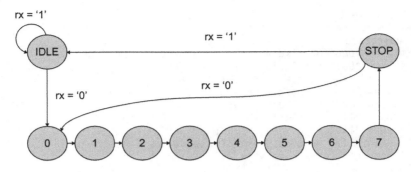

Figure 15.5
Basic serial data receiver state machine.

We can implement this simple state machine in VHDL using the following model:

```
1    library ieee;
2    use ieee.std_logic_1164.all;
3    use ieee.std_logic_unsigned.all;
4
5    entity serialrx is
6      port(
7        clk : in std_logic;
8        rst : in std_logic;
9        rx : in std_logic;
10         dout : out std_logic_vector (7 downto 0)
11         );
12   end serialrx;
13
14   architecture simple of serialrx is
15     type state is (idle, s0, s1, s2, s3, s4, s5, s6, s7, stop);
16     signal current_state, next_state : state;
17     signal databuffer : std_logic_vector(7 downto 0);
18   begin
19   receive: process (rst, clk)
20   begin
21   if rst='0' then
22       current_state <= idle;
23       for i in 7 downto 0 loop
24         dout(i) <= '0';
25       end loop;
26   elsif rising_edge(clk) then
27
28         case current_state is
29         when idle =>
30           if rx = '0' then
31               next_state <= s0;
32           else
33               next_state <= idle;
34           end if;
35     when s0 =>
36     next_state <= s1;
```

```
37      databuffer(0) <= rx;
38        when s1 =>
39      next_state <= s2;
40      databuffer(1) <= rx;
41        when s2 =>
42      next_state <= s3;
43      databuffer(2) <= rx;
44        when s3 =>
45      next_state <= s4;
46      databuffer(3) <= rx;
47        when s4 =>
48      next_state <= s5;
49      databuffer(4) <= rx;
50        when s5 =>
51      next_state <= s6;
52      databuffer(5) <= rx;
53        when s6 =>
54      next_state <= s7;
55      databuffer(6) <= rx;
56        when s7 =>
57      next_state <= stop;
58      databuffer(7) <= rx;
59          when stop =>
60             if rx = '0' then
61                 next_state <= s0;
62             else
63                 next_state <= idle;
64             end if;
65          dout <= databuffer;
66      end case;
67      current_state <= next_state;
68      end if;
69      end process;
70      end;
```

In turn, the same approach of a state machine can be used in Verilog, with the model as shown below:

```verilog
1      module serial_rx (
2      dout, // Output Value
3      clk, // Clock
4      rst, // Reset
5      rx // Input receiver value
6      );
7
8
9      output [3:0] dout;
10     input clk;
11     input rst;
12     input rx;
13
14     reg [3:0] dout;
15
16     reg [3:0] current_state, next_state; // state variable
17
```

```
18    parameter idle=0,s0=1, s1=2, s2=3, s3=4,s4=5,s5=6,s6=7,s7=8,stop=9;
19
20    always @(state)
21      begin
22        case (state)
23        s0:
24          dout[0] = rx;
25        s1:
26          dout[1] = rx;
27        s2:
28          dout[2] = rx;
29        s3:
30          dout[3] = rx;
31        s4:
32          dout[4] = rx;
33        s5:
34          dout[5] = rx;
35        s6:
36          dout[6] = rx;
37        s7:
38          dout[7] = rx;
39      endcase
40    end
41
42    always @(posedge clk)
43    begin
44      if (rst == 0)
45        state = idle;
46      else
47        case (state)
48          idle:
49        state = s0;
50          s0:
51        state = s1;
52          s1:
53        state = s2;
54          s2:
55            state = s3;
56          s3:
57        state = s4;
58          s4:
59        state = s5;
60          s5:
61        state = s6;
62          s7:
63        state = stop;
64          stop:
65        state = idle;
66          default:
67            state=idle;
68        endcase
69    end
70
71    endmodule
```

15.8 Universal Serial Bus

The Universal Serial Bus (USB) protocol has become pervasive and ubiquitous in the computing and electronics industries in recent years. The protocol supports a variety of data rates from low speed (10 kbits/s to 100 kbits/s) up to high speed devices (up to 400 Mbits/s). While in principle it is possible to create FPGA interfaces directly to a USB bus, for anything other than the lower data rates it requires accurate voltage matching and impedance matching of the serial bus. For example, the low data rates require 2.8 V (1) and 0.3 V (0), differentially, whereas the high speed bus requires 400 mV signals, and in both cases termination resistors are required.

In practice, therefore, it is common when working with FPGAs to use a simple interface chip that handles all the analog interface issues and can then be connected directly to the FPGA with a simple UART style interface. An example device is the Silicon Labs CP2101, that takes the basic USB Connector pins (Differential Data and Power & Ground) and then sets up the basic serial data transmission pins. The block diagram of this device is given in Figure 15.6.

The pins on this device are relatively self explanatory and are summarized in the following table.

nRST	The Reset pin for the Device Active Low
Suspend	This pin shows when the USB device is in SUSPEND mode Active High
nSuspend	The Active Low (i.e., inverse) of the SUSPEND pin
RI	Ring Indicator
DCD	Data Carrier Detection shows that data is on the USB line Active Low
DTR	Data Transmit Detection this is active low when the line is ready for data transmission
TXD	Asynchronous Data transmission line
RXD	Asynchronous Data received line
RTS	Clear to Receive Active Low
CTS	Clear to Send Active Low

The basic operation of the serial port starts from the use of the TXD and RXD (data) lines. If the configuration is as a NULL modem with no handshaking, it is possible to simply use the transmit (TXD) and receive (RXD) lines alone.

If you wish to check that the line is clear for sending data, then the RTS signal can be set (Request to Send), in this case active low, and if the line is ready, then the CTS line will go low and the data can be sent. This basic scheme is defined in such a way that, once the receiver signal goes low, the transmitter can send at any rate, the assumption being that the receiver can handle whatever rate is provided. The protocol can be made more capable by using the DTR line, and this notifies the other end of the link that the device is ready for receiving data communications. The Data Carrier Detection (DCD) line is not used directly in the link, but indicates that there is a valid communications link between the devices. We can develop a

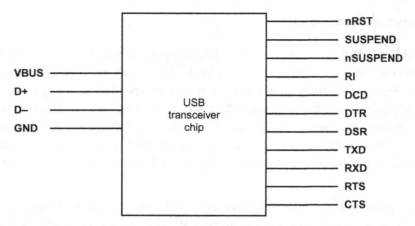

Figure 15.6
USB transceiver chip CP2101.

VHDL model for such a communications link with as much complexity as we need to communicate with the hardware in the system under consideration, starting with a simple template:

```
1     Entity serial_handler is
2       Port(
3         Clk : in std_logic;
4         Nrst : in std_logic;
5         Data_in : in std_logic;
6         Data_out : out std_logic;
7         TXD : out std_logic;
8         RXD : in std_logic
9         );
10    End entity serial_handler;
```

In this initial model, we have a simple clock and reset, with two data connections for the synchronous side, and the TXD and RXD asynchronous data communications lines. We can put together a simple architecture that simply samples the data lines and transfers them into an intermediate variable for use on the synchronous side of the model:

```
1     architecture basic of serial_handler is
2     begin
3       p1 : process (clk )
4       begin
5         if rising_edge(clk) then
6           rxd_int <= rxd;
7         end if;
8       end process p1;
9     end architecture basic;
```

We can extend this model to handle the transmit side also, using a similar approach:

```
1    Architecture mod of serial_handler is
2    Begin
3     P1 : process (clk )
4     Begin
5      If rising_edge(clk) then
6        Data_out <= rxd;
7        Txd <= data_in;
8      End if;
9     End process p1;
10   End architecture mod;
```

This entity is the equivalent to a NULL modem architecture. If we wish to add the DTR notification that the device is ready for receiving data, we can add this to the entity list of ports and then gate the receive data if statement using the DTR signal:

```
1    entity serial_handler is
2     port(
3      clk : in std_logic;
4      nrst : in std_logic;
5      data_in : in std_logic;
6      data_out : out std_logic;
7      dtr : in std_logic;
8      txd : out std_logic;
9      rxd : in std_logic
10     );
11   end entity serial_handler;
12   architecture serial_dtr of serial_handler is
13   begin
14     p1 : process (clk )
15     begin
16      if rising_edge(clk) then
17        if dtr = 0 then
18          data_out <= rxd;
19        end if;
20        txd <= data_in;
21      end if;
22     end process p1;
23    end architecture basic;
```

Using this type of approach we can extend the serial handler to incorporate as much or as little of the modem communications link protocol as we require.

In a similar manner we can generate a similar Verilog serial handler module and then add in the same basic behavior of the handler as required.

```
1    module serial_handler (
2    clk, // clock
3    nrst, // reset
4    datain, // data in
5    dataout, // data out
6    txd, // Tx out
7    rxd  //Received data in
8    );
```

As before, we have a simple clock and reset, with two data connections for the synchronous side, and the TXD and RXD asynchronous data communications lines. We can put together a simple module process that samples the data lines and transfers them into an intermediate variable for use on the synchronous side of the model:

```
1    always @ (posedge clk)
2    begin
3      rxd_int <= rxd;
4    end
```

15.9 Summary

In this chapter, we have introduced a variety of serial communications coding and decoding schemes, and also reviewed the practical methods of interfacing using RS-232 and a USB device. Clearly, there are many more variations on this theme, and in fact a complete USB handler description would be worthy of a book in itself.

PART 4

Optimizing Designs

In this part of the book we will introduce a number of "advanced" topics. In the other parts of the book, the emphasis is on the "what" but in this part it is more about the "how." How can we make designs synthesize? How can our designs be made smaller or faster? How can we interface to mixed signal systems in practice? How can we develop verifiable designs? All of these design challenges will be addressed in this part of the book.

CHAPTER 16

Design Optimization

16.1 Introduction

The area of design optimization is where the performance of a design can be made drastically better than an initial naive implementation. Before discussing details of how to make the designs optimal for the individual goals of speed, area and power (the "big three" for design optimization generally in digital design and particularly for FPGAs), it is useful to discuss some principles of what happens when we synthesize a function into hardware.

There are two main areas for optimization of the design when working with FPGAs. The first is in the optimization of the RTL code, which is leading to an optimal description of the design hardware in terms of logic expressions. The second key area is in the basic logic minimization prior to the mapping of low level functions to the individual technology gates.

16.2 Techniques for Logic Optimization

There are two approaches to minimizing the logic in a design, one that maintains the hierarchy and the other that flattens it. Often a synthesis tool will allow the user to choose which option is required. Clearly the advantage of flattening a design is that the logic can be considered as a whole, whereas if the logic hierarchy is maintained, then there may be structural aspects of the design that will be of benefit to the behavior of the circuit as a whole.

The basic approach of the logic minimization is to reduce the logic equation set to a two level form (otherwise known as sum-of-products). The most common approach for simple designs is to use a Karnaugh map to show the input and output variables graphically and then produce an output expression that can provide the same outputs but using a smaller amount of logic than the original Boolean expressions.

For example, consider the basic 4 input Karnaugh map shown in Figure 16.1.

When a logic expression is described using a logic equation, we can select all valid outputs by circling all the required output 1s and this defines the basic logic behavior. The basic technique is to make the circles as large as possible to encompass as many output 1s with as few input variables as possible. For example, if a basic logic equation was defined as

Design Recipes for FPGAs. http://dx.doi.org/10.1016/B978-0-08-097129-2.00016-7

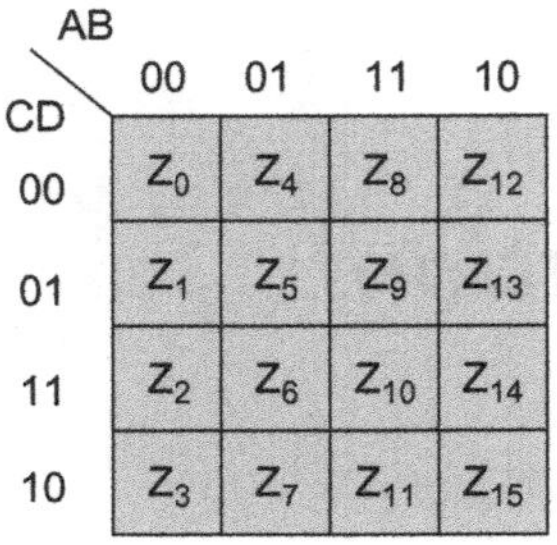

Figure 16.1
Basic 4 input Karnaugh map.

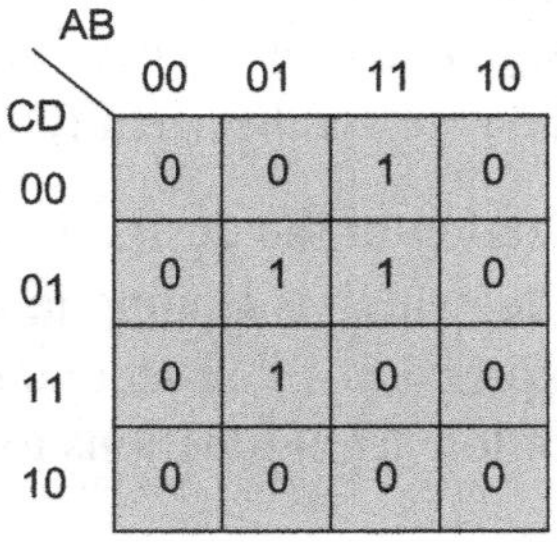

Figure 16.2
Specific Karnaugh map example.

$$Z = AB\overline{CD} + B\overline{C}D + \overline{A}BCD \tag{16.1}$$

then the resulting Karnaugh map would be as shown in Figures 16.2 and 16.3.

Currently, with this basic implementation this would require three, 3 input AND gates, a 3 input OR gate and several inverters. We can see from the Karnaugh map, however, that if we define only two of those logic functions, then there is redundancy in the original definition, and we can reduce this to the same output for two logic combinations of the input.

We could therefore define this model using the simplified expression given in Equation (16.2).

$$Z = A.B.\overline{C} + \overline{A}.B.D \tag{16.2}$$

This has clearly reduced the size of the logic by one 3 input AND gate and the OR gate has reduced to a 2 input gate.

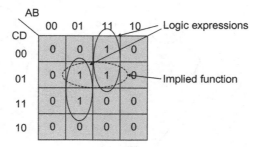

Figure 16.3
Karnaugh Map functions.

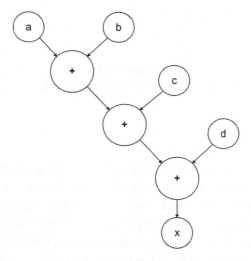

Figure 16.4
Naive dataflow diagram for addition.

16.3 *Improving Performance*

Consider a simple example of an addition $X = A + B + C + D$, where all the variables are digital words. We could implement this using adders, taking two numbers at a time and then adding the answer to the next input. This would give the data flow diagram shown in Figure 16.4.

This implementation requires three adders and takes three cycles to get the answer. If we were more systematic with the same resources, we could reduce this to two cycles by adopting a different structure, as shown in Figure 16.5.

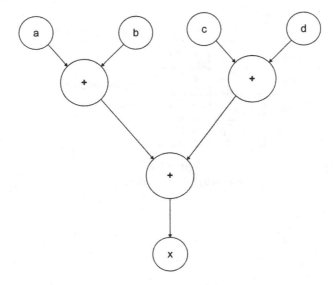

Figure 16.5
Reduced cycle implementation.

This is a classic case of an expression tree being reduced so that the control path can take fewer cycles, but achieving the same data path result. We could also the envisage the case where we only use a single addition block, but use registers to store the intermediate sums and then pipeline the sums until we complete the expression. This would potentially take the longest, but would result in the smallest area requirement as there would only be the need for a single addition block (of course, this would be a trade-off with an increased number of registers).

16.4 Critical Path Analysis

Another approach to logic optimization is to analyze the critical path through a design from a timing perspective. This is often carried out automatically by the synthesis software; for example, the Synopsys© Design Compiler software automatically generates a synthesized schematic that highlights the critical path through the design for timing and as such designers can concentrate their efforts on that area of the design to improve the overall throughput in that case (see Figure 16.6).

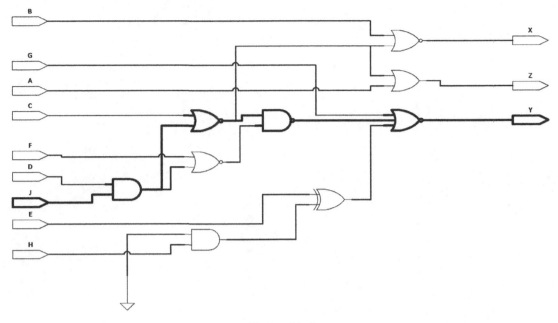

Figure 16.6
Reduced cycle implementation.

16.5 Summary

This chapter has discussed some techniques for improving the performance of designs using FPGAs and how they work. Much of the actual optimization is taken care of in the synthesis software; however, it is useful to understand the processes involved so if a specific target is required for optimization, this can be achieved in a reasonable time in a controlled manner.

CHAPTER 17

Behavioral Modeling in using HDLs

17.1 Introduction

There is a real need to abstract to a higher level in many designs to make the overall system level design easier. There is less need to worry about details of implementation at the system level if the design can be expressed behaviorally, especially if the synthesis method can handle any clock, partitioning, or implementation issues automatically.

Furthermore, by using system level, or behavioral, analysis, decisions can be made early in the design process so that potentially costly mistakes can be avoided. Preliminary area and power estimates can be made and key performance specifications and architectural decisions can be made using this approach, without requiring to have detailed designs for every block.

17.2 How to Go from RTL to Behavioral HDL Descriptions

The abstraction from an RTL (Register Transfer Level) hardware description language (HDL) to behavioral is straightforward in one sense, in that the resulting HDL (whether Verilog or VHDL) is actually simpler. There is no need to ensure that correct clocking takes place, or that separate processes are implemented for different areas of the architecture, or even separate components instantiated.

It is useful to consider an example to illustrate this point by looking at the difference between the RTL and behavioral HDL in an example such as a cross product multiplier. In this case we will demonstrate the RTL method and then show how to abstract to a behavioral model. First, consider the specification for the model in Figure 17.1, which has the data path model as shown in Figure 17.2.

17.3 Implementing the Behavioral Model using VHDL

The first task is to define the types for the VHDL for the entity of the model and this is shown in the following code. Notice that we have defined a new type, sig8, that is a signed type and a vector based on this for the cross product multiplications.

```
1    library ieee;
2    use ieee.std_logic_1164.all;
```

Design Recipes for FPGAs. http://dx.doi.org/10.1016/B978-0-08-097129-2.00017-9

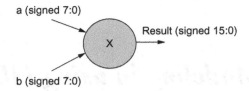

Figure 17.1
Cross product multiplier specification.

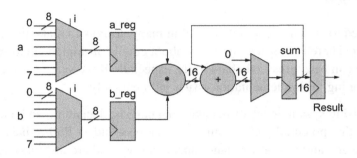

Figure 17.2
Cross product multiplier data path model.

```
3    use ieee.numeric_std.all;
4    package cross_product_types is
5      subtype sig8 is signed (7 downto 0);
6      type sig8_vector is array
7        (natural range<>) of sig8;
8    end package;
```

The entity can now be put together and is shown as follows. Notice that for RTL we require both a clock and a reset.

```
1    library ieee;
2    use ieee.std_logic_1164.all;
3    use ieee.numeric_std.all;
4    use work.cross_product_types.all;
5
6    entity cross_product is
7      port(
8        a,b : in sig8_vector(0 to 7);
9        clk, reset : in bit;
10       result : out signed(15 downto 0)
11     );
12   end entity cross_product;
```

The basic architecture can be set up with the basic internal signals defined, and the processes will be explained separately.

```
1    architecture rtl of cross_product is
2      signal I : unsigned ( 2 downto 0);
3      signal ai, bi : sig8;
4      signal product, add_in, sum, accumulator : signed (15 downto 0);
5    begin
6    control: process (clk)
7    begin
8      if clkevent and clk = 1 then
9        if reset = 1 then
10         i <= (others => 0);
11       else
12         i <= i + 1;
13       end if;
14     end if;
15   end process;
16   a_mux: ai <= a(i);
17   b_mux <= bi <= b(i);
18   multiply: product <=ai * bi;
19   z_mux: add_in <= X 000 when i = 0 else accumulator;
20
21   accumulate: process (clk)
22   begin
23     if clkevent and clk = 1 then
24       accumulator <= sum;
25     end if;
26   end process;
27
28   output : result <= accumulator;
29   end;
```

Notice that there are two processes, one for the accumulation and the other to handle the multiplication. One important aspect is that it is not immediately obvious what is going on. Even in this simple model it is difficult to extract the key behavior of the state machine. In a complex controller it verges on the impossible unless the structure is well known and understood, which is an important lesson when using any kind of synthesis tool using VHDL or Verilog at any level.

Now reconsider using behavioral VHDL instead. The model uses the same packages and libraries as the RTL model; however, notice that there is no need for an explicit clock or reset.

```
1    library ieee;
2    use ieee.std_logic_1164.all;
3    use ieee.numeric_std.all;
4    use work.cross_product_types.all;
5
6    entity cross_product is
7      port(
8        a,b : in sig8_vector(0 to 7);
9        result : out signed(15 downto 0)
10     );
11   end entity cross_product;
```

In this model, the architecture becomes much simpler and can be modeled in a much more direct way than the RTL approach.

```
1    architecture behav of cross_product is
2    begin
3
4      process
5        variable sum : signed(15 downto 0);
6      begin
7        sum := to_signed(0,16);
8        for i in 0 to 7 loop
9          sum := sum + a(i) * b(i);
10       end loop;
11       result <= sum;
12       wait for 100 ns;
13     end process;
14
15   end architecture;
```

Notice that it is much easier to observe the functionality of the model and also the behavior can be debugged more simply than in the RTL model. The design is obvious, the code is readable and the function is easily ascertained. Note that there is no explicit controller; the synthesis mechanism will define the appropriate mechanism. Also notice that the model is defined with a single process. The synthesis mechanism will partition the design depending on the optimization constraints specified.

Note the wait statement. This introduces an implicit clock delay into the system. Obviously this will depend on the clock mechanism used in reality. There is also an implied reset. If an explicit clock is required then use a wait until rising_edge(clk) or similar approach, while retaining the behavioral nature of the model.

17.4 Implementing the Behavioral Model using Verilog

As before, the model (in this case a Verilog module) can now be put together and is shown here. Notice that for RTL we require both a clock and a reset.

```
1    module cross_product (
2      clk, // clock
3      rst, // reset
4      a,   // number a
5      b,   // number b,
6      result // result of the product
7      );
8
9    input clk;
10   input rst;
11
12   input signed [7:0] a;
13   input signed [7:0] b;
14
```

```
15    output reg [15:0] result;
16
17    reg [2:0] i;
18
19    always @ (posedge clk)
20    begin
21      if (rst = 1 ) then
22        i = 3b'000;
23      else
24        i = i + 1;
25      end if
26      accumulator <= sum;
27
28      if (i=0) then
29        addin <= 0;
30      else
31        addin <= accumulator;
32      end if
33    end
34
35    ai <= a[i];
36    bi <= b[i];
37    multiply <= ai * bi;
38    result <= accumulator;
39
40    endmodule
```

Again, even with Verilog which generally has a little simpler syntax than VHDL, in this simple model it is difficult to extract the key behavior of the state machine. In a complex controller it verges on the impossible unless the structure is well known and understood, which is an important lesson when using any kind of synthesis tool using VHDL or Verilog at any level.

Now reconsider using behavioral code instead. The model uses the same packages and libraries as the RTL model; however, notice that there is no need for an explicit clock or reset.

```
1     module cross_product (
2       a,    // number a
3       b,    // number b,
4       result // result of the product
5       );
6
7     input clk;
8     input rst;
9
10    input signed [7:0] a;
11    input signed [7:0] b;
12
13    output reg [15:0] result;
14
15    reg [2:0] i;
16
17    always @ (a or b)
18    begin
19      for (i = 0; i < 8; i = i +1) begin
```

```
20      begin
21        ai <= a[i];
22        bi <= b[i];
23        accumulator <= accumulator + ai * bi;
24      end
25    end
26
27    result <= accumulator;
28
29    endmodule
```

Notice that it is much easier to observe the functionality of the model and also the behavior can be debugged more simply than in the RTL model. The design is obvious, the code is readable and the function is easily ascertained. Note that there is no explicit controller, as the synthesis mechanism will define the appropriate mechanism. Also notice that the model is defined with a single module. The synthesis mechanism will partition the design depending on the optimization constraints specified. This is easily parameterized, modified and clear.

17.5 Summary

Behavioral modeling is a useful technique for both initial design ideas and also as the starting point for an RTL design. It is important to remember, however, that quite a lot of behavioral HDL cannot be synthesized and is therefore purely for conceptual design or use in test benches. In order to make this a practically useful design tool, the designer can take advantage of the ability of VHDL to have numerous architectures, or Verilog to have numerous submodules, and, by using the same test bench, validate the RTL against the behavioral model to ensure correctness.

In summary, we can use behavioral modeling early with high impact to:

• Carry out fast functional simulation;
• Make performance criteria/ Design trade-offs;
• Investigate nonlinear effects;
• Look at implementation issues;
• Carry out topology evaluation.

CHAPTER 18

Mixed Signal Modeling

18.1 Introduction

With the increasingly high level of system integration it is becoming necessary to model not only electronic behavior of systems, but also interfaces to "real world" applications and the detailed physical behavior of elements of the system in question. The emergence of standard languages such as VHDL-AMS, Verilog-AMS, and Verilog-A has made it possible to now describe a variety of physical systems using a single design approach and simulate a complete system. Application areas where this is becoming increasingly important include mixed-signal electronics, electromagnetic interfaces, integrated thermal modeling, electromechanical and mechanical systems (including MEMS), fluidics (including hydraulics and microfluidics), power electronics with digital control, and sensors of various kinds. This is becoming increasingly relevant for systems using digital electronics such as microprocessors and even more so with FPGAs, as they offer the ability to manage multiple interfaces in parallel.

In this chapter, we will show how the behavioral modeling of multiple energy domains is achieved using mixed-signal modeling languages such as VHDL-AMS or Verilog-AMS, demonstrating with the use of examples how the interactions between domains takes place, and providing an insight into design techniques for a variety of these disciplines. The basic framework is described, showing how standard packages can define a coherent basis for a wide range of models, and specific examples are used to illustrate the practical details of such an approach. Examples such as integrated simulation of power electronics systems including electrical, magnetic and thermal effects, mixed-domain electronics, and mechanical systems are presented to demonstrate the key concepts involved in multiple energy domain behavioral modeling.

18.2 Basic Modeling Approach for VHDL-AMS

The basic approach for modeling devices in VHDL-AMS is to define a model entity and architecture(s). The model entity defines the interface of the model to the system and includes connection points and parameters. A number of architectures can be associated with an entity to describe the model behavior, such as a behavioral or physical level description. A complete model consists of a single entity combined with a single architecture. The domain or technology type of the model is defined by the type of terminal used in the entity declaration of the ports. The IEEE Std 1076.1.1 defines standard types for multiple energy domains including electrical, thermal, magnetic, mechanical, and radiant systems. Within the architecture of the model, each energy domain type has a defined set of through and across variables (in the electrical domain these are voltage and current, respectively) that can be used to define the relationship between the model interface pins and the internal behavior of the model.

In the "conventional" electronics arena, the nature of the VHDL-AMS language is designed to support "mixed-signal" systems (containing digital elements, analog elements and the boundary between them) with a focus on IC design. Where the strengths of the VHDL-AMS language have really become apparent, however, is in the multi-disciplinary areas of mechatronic and microelectromechanical systems (MEMS). In this chapter, I have highlighted several interesting examples that illustrate the strengths of this modeling approach, with emphasis on multiple-domain simulations.

18.3 Introduction to VHDL-AMS

VHDL-AMS is a set of analog extensions to standard digital VHDL to allow mixed signal modeling of systems. The VHDL-AMS language was approved as IEEE standard 1076.1 in 1999; however, it is important to note that IEEE 1076.1-1999 encompasses the complete digital VHDL 1076 standard and is not a subset.

The standard does not specify any libraries for analog disciplines, for example, electrical, mechanical, etc. This is a separate exercise and is covered by a subset working group IEEE 1076.1.1, which was released as IEEE Standard 1076.1.1 in 2004.

In order to put the extensions into context it is useful to show the scope of VHDL and then VHDL-AMS alongside it and this is shown in Figure 18.1:

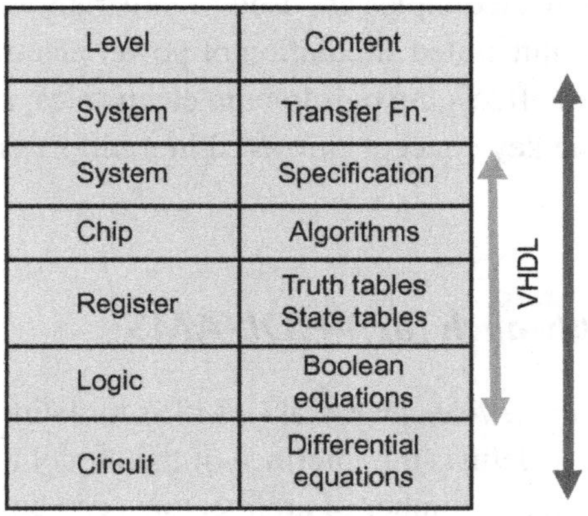

Level	Content
System	Transfer Fn.
System	Specification
Chip	Algorithms
Register	Truth tables State tables
Logic	Boolean equations
Circuit	Differential equations

Figure 18.1
Scope of VHDL-AMS.

The key extension of VHDL-AMS over VHDL is the ability to look upward to transfer functions (behavioral and in the Laplace domain) and downward to differential equations at the circuit level. This gives the possibility for the designer to think about design from a systems perspective (in terms of high level transfer function models) and also a "real world" systems view (linking to different domains).

The specific extensions of VHDL for VHDL-AMS can be summarized as follows:

- A new type of port called TERMINALS—basically analog pins.
- A new type of TYPE called a NATURE, which defines the relationship between analog pins and variables.
- A new type of variable called a QUANTITY, which is an analog variable.
- A new type of variable assignment that is used to define analog equations that are solved simultaneously.
- Differential equation operators for derivative ('DOT) and integration ('INTEG) with respect to time.
- IF statements for equations (IF USE).
- Break statement to initialize the nonlinear solver.
- STEP LIMIT Control for limiting the analog time step in the solver.

18.4 VHDL-AMS Analog Pins: TERMINALS

In order to define analog pins in VHDL-AMS we need to use the TERMINAL keyword in a standard entity PORT declaration. For example, if we have a two pins device that has two analog pins (of type electrical, more on this later), then the entity would have the basic form as shown here:

```
1    library ieee;
2    use ieee.electrical_systems.all;
3    entity model is
4    generic();
5    port(
6      terminal p : electrical;
7      terminal m : electrical
8        );
9    end entity;
```

Notice that as the VHDL-AMS extensions are defined as an IEEE standard, then the use of a standard library such as electrical pins requires the use of the:

```
1    electrical_systems.all;
```

packages from the IEEE library.

Notice that the pins do not have a direction assigned as analog pins are part of a conserved energy system and are therefore solved simultaneously.

18.5 Mixed Domain Modeling

In order to use standard models, there has to be a framework for terminals and variables, which is where the standard packages are used. There is a complete IEEE Std (1076.1.1) that defines the standard packages in their entirety; however, it is useful to look at a simplified package (electrical systems in this case) to see how the package is put together.

For example electrical systems models need to be able to handle several key aspects:

- electrical connection points;
- electrical through variables, that is, current;
- electrical across variables, that is, voltages.

The electrical systems *package* needs to encompass these elements.

First the basic subtypes need to be defined. In all the analog systems and types, the basic underlying VHDL type is always *real* and so the voltage and current must be defined as subtypes of *real*.

```
1       subtype voltage is real;
2       subtype current is real;
```

Notice that there is no automatic unit assignment for either, but this is handled separately by the unit and symbol attributes in IEEE Std 1076.1.1. For example, for voltage the unit is defined as *Volt* and the symbol is defined as *V*.

The remainder of the basic electrical type definition then links these subtypes to the through and across variable of the type, respectively:

```
1     package electrical_system is
2       subtype voltage is real;
3       subtype current is real;
4       nature electrical is
5         voltage across
6         current through
7         ground reference;
8     end package electrical_system;
```

18.6 VHDL-AMS Analog Variables: Quantities

Quantities are purely analog variables and can be defined in one of three ways. Free quantities are simply analog variables that do not have a relationship with a conserved energy system. Branch quantities have a direct relationship between one or more analog terminals, and finally source quantities are used to define special source functions (such as AC sources or Noise sources).

For example, to define a simple analog variable called x, which is a voltage but not related directly to an electrical connection (terminal), then the following VHDL could be used.

```
1    quantity x : voltage;
```

On the other hand, a branch between two electrical pins has a through variable (current) and an across variable (voltage) and this requires a branch quantity so that the complete description can be solved simultaneously. For example, the complete quantity declaration for the voltage (v) and current (i) of a component between two pins (p and m) could be defined as:

```
1    quantity v across i through p to m;
```

18.7 Simultaneous Equations in VHDL-AMS

In VHDL-AMS the equations are analog and solved simultaneously, which is in contrast to signals that are solved concurrently using logic techniques and variables that are evaluated sequentially. For example, if we have two equations that we wish to solve simultaneously:

$$y = x^2 \tag{18.1}$$

$$x = z^3 \tag{18.2}$$

then these must be declared as equations using the == operator in VHDL-AMS, to ensure that they are computed simultaneously and not sequentially.

For example, in VHDL-AMS to solve the first equation, we need to use the == operator:

```
1    y == x**2;
```

where both y and x have to be defined as real numbers (quantities or other VHDL variable types).

18.8 A VHDL-AMS Example: A DC Voltage Source

In order to illustrate some of these basic concepts, consider a simple example of a dc voltage source. This has two electrical pins *p* and *m*, and a single parameter *dc_value* that is used to define the output voltage of the source. The model symbol is shown in Figure 18.2.

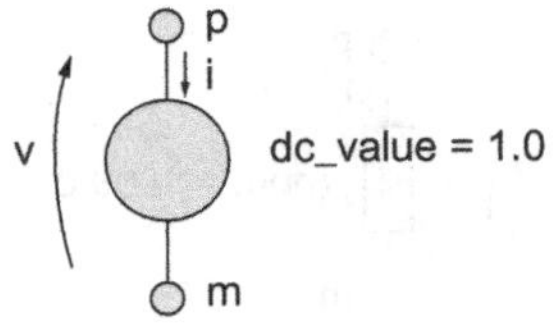

Figure 18.2
Basic VHDL-AMS voltage source.

This can be modeled in VHDL-AMS in two parts, the entity and architecture. First consider the entity. This has two electrical pins, so we need to use the *ieee.electrical_systems.all* package and therefore the ports are to be declared as terminals. Also the generic must be defined as a real number with the default value also defined as a real number (e.g., 1.0).

```
1    library ieee;
2    use ieee.electrical_systems.all;
3    entity v_dc is
4    generic(
5      dc_value : real := 1.0);
6    port(
7      terminal p : electrical;
8      terminal m : electrical
9        );
10   end entity;
```

The architecture must define the quantities for voltage and current through the source and then link those to the terminal pin names. Also, the output equation of the source must be modeled as an analog equation in VHDL-AMS using the == operator to implement the function $v = dc\_value$.

```
1    architecture simple of v_dc is
2      quantity v across i through p to m;
3    begin
4      v == dc_value;
5    end architecture simple;
```

18.9 A VHDL-AMS Example: Resistor

In the case of the resistor, the basic entity is very similar to the voltage source with two electrical pins p and m with a single generic, this time for the nominal resistance rnom Figure 18.3.

This can be modeled in VHDL-AMS in two parts, the entity and architecture. First consider the entity. This has two electrical pins, so we need to use the ieee.electrical_systems.all; package and therefore the ports are to be declared as TERMINALS. Also, the generic rnom must be defined as a real number, with the default value also defined as a real number (e.g., 1000.0).

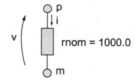

Figure 18.3
Basic VHDL-AMS resistor.

```
1    library ieee;
2    use ieee.electrical_systems.all;
3    entity resistor is
4    generic(
5      rnom : real := 1000.0);
6    port(
7      terminal p : electrical;
8      terminal m : electrical
9        );
10   end entity;
```

The architecture must define the quantities for voltage and current through the resistor and then link those to the terminal pin names. Also, the output equation of the resistor must be modeled as an analog equation in VHDL-AMS using the == operator to implement the function *v = i * rnom*.

```
1    library ieee;
2    use ieee.electrical_systems.all;
3    entity resistor is
4    generic(
5      rnom : real := 1000.0);
6    port(
7      terminal p : electrical;
8      terminal m : electrical
9        );
10   end entity;
```

18.10 Differential Equations in VHDL-AMS

VHDL-AMS also allows the modeling of linear differential equations using the two differential operators:

- 'dot (Differentiate the variable with respect to time);
- 'integ (Integrate the variable with respect to time).

We can illustrate this by taking two examples, a capacitor and an inductor. First consider the basic equation of a capacitor:

$$i = C\frac{\mathrm{d}V}{\mathrm{d}t} \tag{18.3}$$

Using a similar model structure as the resistor, we can define a model entity and architecture, but what about the equation? In VHDL-AMS, the 'dot operator is used on the voltage to represent the differentiation as follows:

```
1    i == c*v'dot;
```

Therefore, a complete capacitor model in VHDL-AMS could be implemented as follows:

```
1    library ieee;
2    use ieee.electrical_systems.all;
3    entity capacitor is
4    generic(
5      cap : real := 1.0e-9);
6    port(
7      terminal p : electrical;
8      terminal m : electrical
9        );
10   end entity;
11
12   architecture simple of capacitor is
13     quantity v across i through p to m;
14   begin
15     i == cap * v'dot;
16   end architecture simple;
```

What about an inductor? The basic equation for an inductor is given as:

$$i = 1/L \int v \, dt \qquad (18.4)$$

which could also be written as:

$$v = L\frac{di}{dt} \qquad (18.5)$$

Obviously, the most direct way to implement this equation would be to use the *'integ* operator; however, care should be taken with the integration operator as some simulators do not handle the integration function in the same manner (in fact, some simulators do not support it at all well). Obviously the initial condition must be considered and in addition different implementations can occur across simulators. One standard approach is to use what is called *implicit integration*, whereby, using the differential equation, the integral function can be inferred. However, the resulting implementation in its simplest form could be as follows:

```
1    library ieee;
2    use ieee.electrical_systems.all;
3    entity inductor is
4    generic(
5      ind : real := 1.0e-9);
6    port(
7      terminal p : electrical;
8      terminal m : electrical
9        );
10   end entity;
11
12   architecture simple of inductor is
13     quantity v across i through p to m;
14   begin
15     i == (1.0/ind) * v'integ;
16   end architecture simple;
```

18.11 Mixed-Signal Modeling with VHDL-AMS

Most design engineers are familiar with the concepts of digital or analog modeling; however, a true understanding of true mixed-signal modeling is often lacking. In order to explain the term *mixed-signal modeling* it is necessary to review what we mean by analog and digital modeling first. First, consider digital modeling techniques.

Digital systems can be modeled using digital gates or events. This is a fast way of simulating digital systems structurally and is based on VHDL or Verilog gate level models. Digital simulation with digital computers relies on an event-based approach, so rather than solve differential equations, events are scheduled at certain points in time, with discrete changes in level. The resolution of multiple events and connections is achieved using logical methods. The digital models are usually gates, or logic based, and the resulting simulation waveforms are of fixed, predefined levels (such as 0 or 1). Also, instantaneous changes can take place, that is, the state can change from 0 to 1 with zero risetime.

In the analog world, in contrast, the lowest level of detail in practical electrical system design is the use of analog equation models in an analog simulator; the benchmark of this approach is historically the SPICE simulator. In many cases the circuit is extracted in the form of a netlist. The netlist is a list of the components in the design, their connection points and any parameters (such as length, width, or scaling) that customize the individual devices.

Each device is modeled using nonlinear differential equations that must be solved using a Newton-Raphson type approach. This approach can be very accurate, but is also fraught with problems such as:

- Convergence: If the model does not converge, then the simulation will not give any meaningful result or fail altogether.
- Oscillation: If there are discontinuities, the solution may be impossible to find.
- Time: The simulations can take hours to complete, days for large designs with detailed device models.

In the analog domain the Newton-Raphson approach is generally used to find a solution that relies on calculating the derivatives as well as the function value to obtain the next solution. The basic Newton-Raphson method for nonlinear equations is defined as:

$$x_{n+1} = x_n - \frac{f(x_n)}{f'(x_n)} \qquad (18.6)$$

$F(x_n)$ and $F'(x_n)$ must be explicitly known and coded into the simulator (for SPICE) and this gives an approximate solution to the exact problem. For VHDL-AMS simulators the derivatives must be estimated using a Secant method (or similar).

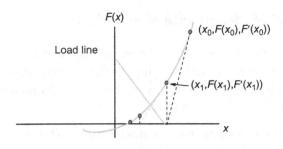

Figure 18.4
Newton-Raphson method.

So given these diametrically opposed methods, how can we put them together? What about mixed signal systems? In these cases, there is a mixture of continuous analog variables and digital events. The models need to be able to represent the boundaries and transitions between these different domains effectively and efficiently. The basic mechanism for checking if an analog variable crosses a threshold is to use the 'above operator in VHDL-AMS.

For example, to check if a voltage *vin* is above 1.0 V, the following VHDL-AMS could be used:

```
1    if ( vin'above(1.0) ) then
2      flag <= true;
3    end if;
```

This can be extended to use parameters in the model, say a threshold voltage parameter *vth* defined previously as a generic or constant.

```
1    if ( vin'above(vth) ) then
2      flag <= true;
3    end if;
```

Notice that the flag is a signal and is therefore able to be used in the sensitivity list to a process enabling digital behavior to be triggered when the threshold is crossed. If the opposite condition is required, that is, below the threshold, then the condition is simply inverted using the *not* operator:

```
1    if ( not vin'above(vth) ) then
2      flag <= true;
3    end if;
```

The digital to analog interface is slightly more complex than the analog to digital interface, inasmuch as the output variable needs to be controlled in the analog domain. When a digital event changes (this can be easily monitored by a sensitivity list in a process), the analog variable needs to have the correct value and the correct rate of change. To achieve this we use the RAMP attribute in VHDL-AMS. Consider a simple example of a digital logic to analog voltage interface.

- when din = '1' vout = 5V
- when din = '0' vout = 0V

This can be implemented using VHDL-AMS as follows:

```
1    process (din) :
2    begin
3      if ( din = '1' ) then
4        vdin = 5.0;
5      else
6        vdin = 0.0;
7      end if;
8    end process;
9    vout == vdin;
```

Clearly there will be problems with this simplistic interface as the transition of vout will be instantaneous, causing potential convergence problems. The technique to solve this problem is to introduce a ramp on the definition of the value of vout with a transition time to change continuously from one value to another:

```
1    vout == dvin'ramp(tt)
```

where tt (the transition time) is defined as a real number, for example, tt : real := 1.0e−9.

An alternative to the specific transition time definition is to limit the slew rate using the SLEW operator. The technique to solve this problem is to introduce a slew rate definition on the definition of the value of vout with a transition time to change continuously from one value to another:

```
1    vout == dvin'slew(max_slew_rate)
```

where max_slew_rate is defined as a real number, for example, max_slew_rate : real := 1.0e6.

18.12 A Basic Switch Model

Consider a simple digitally controlled switch that has the following characteristics:

1. Digital control input (d)
2. Two electrical terminals (p and m)
3. On resistance (Ron)
4. Off resistance (Roff)
5. Turn on time (Ton)
6. Turn off time (Toff)

Using this simple outline a basic switch model can be created in VHDL-AMS. The entity is given here:

```
1    use ieee.electrical_system.all;
2    use ieee.std_logic_1164.all;
3    entity switch is
```

```
4        generic ( ron : real := 0.1; -- on resistance
5        roff : real := 1.0e6; -- off resistance
6           ton : real := 1.0e-6; -- turn on time
7           toff : real := 1.0e-6); -- turn off time
8        port (
9        d : in std_logic;
10       terminal p,m : electrical);
11    end entity switch;
```

The basic structure of the architecture requires that the voltage and current across the terminals of the switch be dependent on the effective resistance of the switch (reff):

```
1     architecture simple of switch is
2       quantity v across i through p to m;
3       quantity reff : real;
4       signal r_eff : real := roff;
5     begin
6       process (d)
7       begin
8        -- the body of the behavior goes here
9       end;
10
11      i = v / reff;
12    end;
```

The process waits for changes on the input digital signal (d) and schedules a signal r_eff to take the value of the effective resistance (ron or roff) depending on the logic value of the input signal. The VHDL for this functionality is shown here:

```
1     process (d)
2       begin
3        if ( d = '1' ) then
4          r_eff <= ron;
5        else
6          r_eff <= roff;
7        end if;
8       end;
```

When the signal r_eff changes, then this must be linked to the analog quantity reff using the ramp function. Previously we showed how the ramp could define a risetime, but in fact it can also define a falltime. Implementing this in the switch model architecture, we get the following VHDL-AMS:

```
1     reff == r_eff'ramp ( ton, toff );
2     i == v / reff;
```

The complete VHDL-AMS model for the switch architecture is given as:

```
1     architecture simple of switch is
2       quantity v across i through p to m;
3       quantity reff : real;
4       signal r_eff : real := roff;
5     begin
6     process (d)
7       begin
8        if ( d = 1 ) then
```

```
9          r_eff <= ron;
10       else
11          r_eff <= roff;
12       end if;
13     end process;
14
15     reff == r_efframp ( ton, toff );
16     i == v / reff;
17   end;
```

18.13 Basic VHDL-AMS Comparator Model

Consider a simple comparator that has two electrical inputs (p and m), an electrical ground (gnd) and a digital output (d). The comparator has a digital output of 1 when p is greater than m and 0 otherwise (Figure 18.5).

The entity defines the terminals (p, m, gnd), digital output (d), input hysteresis (hys), and the propagation delay (td).

```
1    use ieee.electrical_system.all;
2    use ieee.std_logic_1164.all;
3    entity comparator is
4      generic (
5        td : time := 10 ns;
6        hys : real := 1.0e-6;
7        );
8      port (
9        d : out std_logic := '0';
10       terminal p,m,gnd : electrical
11       );
12   end entity comparator;
```

The first step in the architecture is to define the input voltage and basic process structure:

```
1    architecture simple of comparator is
2      quantity vin across p to m;
3    begin
4      p1 : process
5        constant vh : real := abs(hys)/2.0;
6        constant vl : real := -abs(hys)/2.0;
7      begin
8        -- The comparator digital handler goes here
9        wait on vin'above(vh), vin'above(vl);
10     end process;
11   end architecture simple;
```

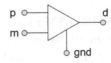

Figure 18.5
Basic VHDL-AMS comparator.

The quantity vin is defined as the voltage across the input pins p and m.

```
1      quantity vin across p to m;
```

Notice that no current is defined (i.e., assumed to be zero) so there is no input current to the comparator. Also notice that there is no input voltage offset defined; this could be added as a refinement to the model later. The process defines the upper and lower thresholds (vh and vl) based on the hysteresis:

```
1      constant vh : real := abs(hys)/2.0;
2      constant vl : real := -abs(hys)/2.0;
```

The process then defines a wait statement, checking vin for crossing either of those threshold values:

```
1      wait on vin'above(vh), vin'above(vl);
```

The final part of the process is to add the digital output logic state dependent on the threshold status of vin:

```
1      if vin'above(vh) then
2        d <= '1' after td;
3      elsif not vin'above(vl) then
4        d <= '0' after td;
5      end if;
```

The output state (d) is then scheduled after the delay time defined by td.

The completed architecture is shown as follows:

```
1      architecture simple of comparator is
2        quantity vin across p to m;
3      begin
4       p1 : process
5        constant vh : real := ABS(hys)/2.0;
6        constant vl : real := -ABS(hys)/2.0;
7       begin
8         if vinabove(vh) then
9           d <= '1' after td;
10        elsif not vinabove(vl) then
11          d <= '0' after td;
12        end if;
13        wait on vin'above(vh), vin'above(vl);
14       end process;
15     end architecture simple;
```

18.14 Multiple Domain Modeling

A final significant application area for VHDL-AMS has been the modeling of electromechanical systems, particularly micromachines (or MEMS). Exactly the same principles are used for these devices, with the mechanical domain models defined as required for the mechanical equations. It is worth noting that the mechanical models are divided into rotational (angular velocity and torque) and translational (force and distance) types. A typical simple example of a mixed domain system is a motor, in this case a simple DC motor. Taking the standard motor equations as shown here, it can be seen that the parameter ke links the rotor speed to the electrical domain (back emf) and the parameter kt links the current to the torque.

This is implemented using the VHDL-AMS model shown as follows:

```
1    library ieee;
2    use ieee.electrical_systems.all;
3    use ieee.mechanical_systems.all;
4
5    entity dc_motor is
6        generic (kt : real;
7                 j  : real;
8                 r  : real;
9                 ke : real;
10                d  : real;
11                l  : real);
12       port (terminal p, m : electrical;
13             terminal rotor : rotational_v );
14    end entity dc_motor;
15
16    architecture behav of dc_motor is
17       quantity w across t through rotor
18       to rotational_v_ref;
19       quantity v across i through p to m;
20    begin
21       v == l*i'DOT + i*r + ke*w;
22       t == i*kt — j*w'DOT — d*w;
23    end architecture behav;
```

18.15 *Introduction to Verilog-AMS*

The extensions to Verilog for analog and mixed signal functions are not defined by a single IEEE standard as is the case with VHDL-AMS, and so there are a number of variants, derived from Verilog, but somewhat dependent on the proprietary simulator used.

The basic concepts are essentially the same and so these can be considered as being largely consistent, while detailed syntax may be slightly different depending on which simulator or language "flavor" is used.

The Verilog extensions for analog and mixed signal systems have broadly the same scope as those for VHDL with support for conserved equations and differential equations (sequential and simultaneous). New types of connections are defined for conservative and analog variables, and finally, the ability to convert from analog to digital (and vice versa) is implemented with specific operators in Verilog-AMS.

18.16 Verilog-AMS: Analog ports

In Verilog, ports are defined by name, direction and then finally type, so for example, a digital gate may have an input called d, which is an input and finally it is defined as a bit, or bus type. Analog signals can be defined in a very similar manner; however, there are some important distinctions to be observed depending on which type of analog variable is required. If a conserved variable (for elements such as resistors, capacitors, and other typical circuit components) is required, then the direction *must* be defined as inout. Input and output are used for signal flow type models (such as simple control blocks) where the models are not to be used in a conserved manner.

The type is then defined using standard libraries which define the through and across variables (similar to the standard packages in VHDL-AMS). These are defined in a series of disciplines, with natures (very similar terminology to VHDL-AMS), so for example in the electrical domain, connection points (called "nodes" in Verilog-AMS) have a specific nature (voltage across and current through) that define the discipline (electrical).

Using this approach, the connection points to a model can be defined using those natures directly. For example, take a simple electrical two-port model, with two pins p and m, both of type electrical, the basic module structure will be as follows:

```
1    module model(p, m);
2    inout p,m;
3    electrical p,m;
4    // Main Model Behavior goes here
5    endmodule;
```

18.17 Mixed Domain Modeling in Verilog-AMS

In Verilog-AMS, there is a "Disciplines" package that defines all the standard technologies and their basic natures. This is usually implemented in a single Verilog-AMS code *disciplines.vams* which must be included in the header of models using Verilog-AMS and these disciplines.

The implication is that these natures are all intrinsically real types, but the nature definition will set up the name, units, tolerances, and related natures. For example, the nature for Voltage could be defined as follows:

```
1    // Potential in volts
2    nature Voltage
3    units = "V";
4    access = V;
5    idt_nature = Flux;
6    'ifdef VOLTAGE_ABSTOL
7    abstol = 'VOLTAGE_ABSTOL;
8    'else
```

```
 9      abstol = 1e-6;
10      `endif
11      endnature
```

With each nature defined, then a complete discipline can then be implemented with the through and across variables defined.

```
1      discipline electrical
2      potential Voltage;
3      flow Current;
4      enddiscipline
```

Note that in Verilog-AMS, the "across" variable is called "potential" and the "through" variable is referred to as the "flow."

18.18 Verilog-AMS Analog Variables

The way that Verilog-AMS manages its behavior in the analog domain is to define each equation in terms of a "branch." Real variables can be defined directly to be used as intermediate variables; however, central to the concept is that of equations that define the through and across equations between two nodes.

For example, if two nodes are defined as p and m, respectively, and are both of type electrical (i.e., the same type) then they can be connected via a branch definition of the voltage (across) and current (through) variables.

For the voltage part of the branch, this is obtained using the following technique:

```
1      // Voltage across pins p and m
2      V(p,m)
```

and similarly for the current:

```
1      // Current through pins p and m
2      I(p,m)
```

Using these basic definitions, equations can then be constructed in Verilog-AMS.

18.19 Verilog-AMS Analog Equations

Now that we have a set of analog connections (nodes) with a specific analog type (from disciplines.vams) and also branch definitions for through and across variables, it is now a simple matter to construct the equations to describe the behavior of the model.

For example, consider the equation for a simple resistor:

$$v = i * r \tag{18.7}$$

Now that we have the voltage V(p,m) and current I(p,m), if we assume that the resistance r has been defined as a parameter of type real already, then the operator <+ can be used to create the governing equation for the analog behavior:

```
1    V(p,m) <+ I(p,m)*r
```

In the Verilog modeling scheme, each section of the model is defined using some form of block statement (such as the "always" statement in the digital world) and in the analog equation section, the same approach is used with the "analog" block. Defining an analog block enables the designer to collect all the analog behavior and separate it from digital expressions.

Using this approach, the core of our simple resistor model would then become something like the following:

```
1    analog begin
2      V(p,m) <+ I(p,m)*r
3    end
```

18.20 A Verilog-AMS Example

18.20.1 DC Voltage Source

As we saw with the VHDL-AMS approach, we can create a simple voltage source that has two pins p and m, with a dc value (dcv) using a simple Verilog-AMS model. In some ways the Verilog-AMS model is simpler to implement than its VHDL-AMS equivalent, as there is no separate entity and architecture, just a single module, and this is shown in the listing:

```
1    module vdc(p,m)
2    inout p; // Positive Terminal
3    inout m; // Negative Terminal
4
5    electrical p,m; // Define ports as electrical
6
7    parameter real dcv = 0.0; // define the DC voltage with a default value of 0.0
8
9    analog begin
10     V(p,m) <+ dcv;
11   end
12
13   endmodule
```

18.20.2 Resistor

In the case of the resistor, the basic model is very similar to the voltage source with two electrical pins p and m with a single parameter, this time for the nominal resistance rnom.

```
1    module resistor(p,m)
2    inout p; // Positive Terminal
3    inout m; // Negative Terminal
```

```
4
5    electrical p,m; // Define ports as electrical
6
7    parameter real rnom = 1.0; // define the resistance with a default value of 1.0
8
9    analog begin
10     V(p,m) <+ I(p,m) * rnom;
11   end
12
13   endmodule
```

18.21 Differential Equations in Verilog-AMS

Verilog-AMS also allows the modeling of linear differential equations using the two differential operators:

- ddt (Differentiate the variable with respect to time)
- idt (Integrate the variable with respect to time)

We can illustrate this by taking two examples, a capacitor and an inductor. First, consider the basic equation of a capacitor:

$$i = C\frac{\mathrm{d}V}{\mathrm{d}t} \tag{18.8}$$

Using a similar model structure as the resistor, we can define a model entity and architecture, but what about the equation? In Verilog-AMS, the ddt function is used on the voltage to represent the differentiation as follows:

```
1    I(p,m) <+ c*ddt(V(p,m))
```

Therefore, a complete capacitor model in VHDL-AMS could be implemented as follows:

```
1    module c(p,m)
2    inout p; // Positive Terminal
3    inout m; // Negative Terminal
4
5    electrical p,m; // Define ports as electrical
6
7    // define the capacitance with a default value of 1.0e-6
8    parameter real cap = 1.0e-6;
9
10   analog begin
11     I(p,m) <+ cap*ddt(V(p,m));
12   end
13
14   endmodule
```

What about an inductor? The basic equation for an inductor is given as follows:

$$i = 1/L \int v \, \mathrm{d}t \qquad (18.9)$$

which could also be written as:

$$v = L\frac{\mathrm{d}i}{\mathrm{d}t} \qquad (18.10)$$

Obviously, the most direct way to implement this equation would be to use the $i \, \mathrm{d}t$ operator; however, care should be taken with the integration operator as some simulators do not handle the integration function in the same manner (in fact, some simulators do not support it at all well). Obviously the initial condition must be considered and in addition different implementations can occur across simulators. One standard approach is to use what is called *implicit integration*, whereby, using the differential equation, the integral function can be inferred. However, the resulting implementation using the differential equation in its simplest form could be as follows:

```
1    module l(p,m)
2    inout p; // Positive Terminal
3    inout m; // Negative Terminal
4
5    electrical p,m; // Define ports as electrical
6
7    // define the inductance with a default value of 1.0e−6
8    parameter real ind = 1.0e−6;
9
10   analog begin
11     V(p,m) <+ ind*ddt(I(p,m));
12   end
13
14   endmodule
```

18.22 Mixed Signal Modeling with Verilog-AMS

As we discussed earlier, the issues of convergence, nonlinearity and also mixed-signal boundaries are an important issue and are no less important in Verilog-AMS. As we saw with VHDL-AMS, the first boundary to consider is from analog to digital, and in Verilog-AMS this can be implemented using the *cross* function.

Using this approach, the crossing can be tested and then also whether it is rising or falling; in this example it is looking for the voltage at node x crossing 2, and the 1 denotes a rising crossing.

```
1    always begin
2    @(cross(V(x) − 2 ,1))
```

```
3    q = 1;
4    end
```

In this case, the transition is looking for a falling cross and in this case the logic signal q is set to 0:

```
1    always begin
2    @(cross(V(x) - 2 ,-1))
3    q = 0;
4    end
```

In the opposite direction, the transition function is used to look for changes in logic values and then to use that to set an analog variable, as in this example where the logic input (din) is checked and, if high, sets the voltage on pin p to 3.3 or 0 V if low.

```
1    V(p) <+ transition((din == 1) ? 3.3 : 0.0);
```

Using these two operators, mixed-signal elements such as comparators, ADCs, or DACs can be implemented.

18.23 *Multiple Domain Modeling using Verilog-AMS*

A final significant application area for Verilog-AMS has been the modeling of electromechanical systems, particularly micromachines (or MEMS). Exactly the same principles are used for these devices, with the mechanical domain models defined as required for the mechanical equations. It is worth noting that the mechanical models are divided into rotational (angular velocity and torque) and translational (force and distance) types. A typical simple example of a mixed-domain system is a motor, in this case a simple DC motor. Taking the standard motor equations as shown following, it can be seen that the parameter ke links the rotor speed to the electrical domain (back emf) and the parameter kt links the current to the torque.

This is implemented using the Verilog-AMS model shown here:

```
1    module motor(p, m, w);
2      inout p,m,w;
3      electrical p,m;     // Electrical Connections
4      rotational_omega w;    // Motor Rotational Shaft
5
6      parameter real r = 1.0e-3; // Winding Resistance
7      parameter real l = 10.0e-3; // Winding inductance
8      parameter real d = 0.0;   // Motor Friction Loss
9      parameter real j = 10.0e-6; // Motor Shaft Inertia
10     parameter real ke = 1.0; // Motor Electrical Constant
11     parameter real kt = 1.0; // Motor Torque Constant
12
13       analog begin
14         // Electrical Equation
15       V(p,m) <+ ke*Omega(w) + r*I(p,m) + l*ddt(I(p,m));
16       // mechanical Equation
```

```
17        Tau(w) <+ kt*I(p,m) - d*Omega(w) - j*ddt(Omega(w));
18        end
19      endmodule
```

18.24 Summary

It has become crucial for effective design of integrated systems, whether on a macro- or microscopic scale, to accurately predict the behavior of such systems prior to manufacture. Whether it is ensuring that sensors or actuators operate correctly, or integrated components such as magnetics also operate correctly, or analyzing the effect of parasitics and non-ideal effects such as temperature, losses, and nonlinearities, the requirement for multiple-domain modeling has never been greater.

Now languages such as VHDL-AMS and Verilog-AMS offer an effective and efficient route for engineers to describe these systems and effects, with the added benefit of standardization leading to interoperability and model exchange. The challenge for the EDA industry is to provide adequate simulation and particularly modeling tools to support engineering design.

The opportunity for FPGA designers is to take advantage of this huge advance in modeling technology and use it to make sure that digital controllers and designs can operate effectively and robustly in real-world applications.

CHAPTER 19

Design Optimization Example: DES

19.1 Introduction

Elsewhere in this book the basics of design optimization are discussed, but in general these are at an RTL level. The use of behavioral modeling has also been described, but in general the use of high-level behavioral synthesis is still rarely used in practice. In this chapter, the use of behavioral synthesis is investigated as an alternative to create optimal designs rather than using an RTL approach.

This chapter describes the experience of designing a Data Encryption Standard (DES) core in Electronic Code Book (ECB) mode using a high-level behavioral synthesis system. The main objective was to write a high-level language description that was both readable and synthesizable. The secondary objective was to explore the area/delay design space of both single and triple DES. The designs were simulated using both the pre-synthesis (behavioral) and post-synthesis (RTL) VHDL, verifying that the outputs were not only the same, but were the expected outputs defined in the test set.

In this chapter, the high-level code has been written in VHDL as the MOODS software only supports VHDL input; therefore there is not a direct Verilog equivalent.

It should be pointed out that there are now more options for the designer than ever before for high-level modeling, including C; however, behavioral VHDL or Verilog is still relatively straightforward to design at the same time as RTL code, as the simulations are easy to manage (and in fact often the same entity could be used, but with different architectures). System-C has become useful for high-level modeling, particularly when coding is involved. However, the ability for hardware designers to handle more architectural issues in addition to the behavioral model still makes behavioral HDL modeling a useful tool.

19.2 The Data Encryption Standard

The Data Encryption Standard, usually referred to by the acronym DES, is a well-established encryption algorithm which was first standardized by NIST in the 1980s. It is described in detail earlier in this book, in Chapter 10, so only the basic information about the algorithm is presented here.

Design Recipes for FPGAs. http://dx.doi.org/10.1016/B978-0-08-097129-2.00019-2

While DES has largely been superseded by the AES (Advanced Encryption Algorithm) it is now common to find the algorithm being used in triplicate (an algorithm known as Triple-DES or TDES for short). This algorithm uses the same DES core, but uses three passes with different keys. DES was designed to be small and fast, and the algorithm is mainly based on shuffling and substitution. There is very little computation involved, which makes it ideal for hardware implementation.

19.3 MOODS

MOODS (Multiple Objective Optimization in Control and Datapath Synthesis) is a high-level behavioral synthesis suite developed at the University of Southampton. It takes as input high-level behavioral VHDL and transforms this into structural VHDL that is behaviorally equivalent. MOODS uses optimization and design space exploration to obtain suitable RTL designs to meet the designer's constraints and requirements.

An optimizer is used to convert the behavioral VHDL into a form that can be described using a simple dataflow graph (DFG) which allows the control flow to be optimized. This is effectively a state machine that can be easily converted into RTL VHDL. The optimization of this with respect to area can be achieved by sharing data units (such as registers) using multiplexing and with respect to delay by combining data units to reduce the number of clock cycles required.

19.4 Initial Design

19.4.1 Introduction

The overall structure of the DES algorithm is shown in Figure 19.1.

The core algorithm is repeated 16 times with a different *subkey* for each round. These subkeys are 48 bits long and are generated from the original 56-bit key. The algorithm was converted directly to VHDL using a functional decomposition style (i.e., functions were created to represent each equivalent function in DES).

19.4.2 Overall Structure

The first stage in this design was to create an entity and an architecture with the required inputs and outputs and a single process containing the overall algorithm. This resulted in the VHDL outline here:

```
library ieee;
use ieee.std_logic_1164.all;
entity DES is
  port (
```

```
    plaintext : in std_logic_vector(1 to 64);
    key : in std_logic_vector(1 to 64);
    encrypt : in std_logic;
    go : in std_logic;
    ciphertext : out std_logic_vector(1 to 64);
    done : out std_logic
  );
end;

architecture behavior of DES is
  subtype vec56 is std_logic_vector(1 to 56);
  ...
  subtype vec64 is std_logic_vector(1 to 64);
begin
  process
  begin
    wait until go = 1;
    done <= 0;
    wait for 0 ns;
    ciphertext <=
    des_core(plaintext, key_reduce(key), encrypt);
    done <= 1;
  end process;
end;
```

This process is a direct implementation of the main DES routine. The only implementation-specific feature is that the model waits for the signal *go* to be raised before

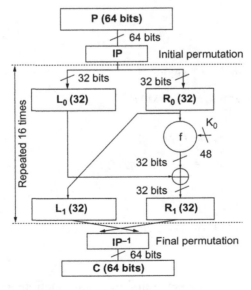

Figure 19.1
Overall structure of the DES algorithm.

starting processing and it raises the signal done at the end of processing, implementing a basic handshaking protocol.

This algorithm requires the two functions key_reduce and des_core. The former strips the parity bits from the key and the latter then implements the whole DES algorithm. The key_reduce function reduces the key from 64 to 56 bits and permutes the bits to form the initial state of the subkey:

```
function key_reduce(key : in vec64) return vec56 is
--moods inline
begin
  return
  key(57) & key(49) & key(41) & key(33) &
  ...
  key(28) & key(20) & key(12) & key(4);
end;
```

The compiler directive –moods inline causes the synthesizer to inline the function. This allows the optimizer more scope for optimization of the circuit. The des_core function applies the basic DES algorithm 16 times on a slice of the data using a different subkey on each iteration:

```
function des_core
  --moods inline
  (plaintext : vec64;
  key : vec56;
  encrypt : std_logic)
return vec64
is
    variable data : vec64;
variable working_key : vec56 := key;
begin
  data := initial_permutation(plaintext);
  for round in 0 to 15 loop
    working_key :=
    key_rotate(working_key,round,encrypt);
    data := data(33 to 64) &
    (f(data(33 to 64),key_compress(working_key))
    xor
    data(1 to 32));
  end loop;
  return
    final_permutation(data(33 to 64) & data(1 to 32));
end;
```

The DES algorithm is made up of the key transformation functions key_rotate and key_compress, and the data transformation functions initial_permutation, f and final_permutation.

19.4.3 Data Transformations

The data transformations initial_permutation and final_permutation are simply hard-wired bit-swapping routines implemented using concatenation.

```
function initial_permutation(data : vec64) return vec64 is
  --moods inline
begin
  return
    data(58) & data(50) & data(42) & data(34) &
    ...
    data(31) & data(23) & data(15) & data(7);
end;

function final_permutation(data : in vec64) return vec64 is
  --moods inline
begin
  return
    data(40) & data(8) & data(48) & data(16) &
    ...
    data(49) & data(17) & data(57) & data(25);
end;
```

The f function is the main data transform, which is applied 16 times to the rightmost half, a 32-bit slice, of the data path. It takes as its second argument a 48-bit subkey generated by the key_compress function.

```
function f(data : vec32; subkey : vec48) return vec32 is
  --moods inline
begin
  return permute(substitute(expand(data) xor
  subkey));
end;
```

The function first takes the 32-bit slice of the datapath and expands it into 48 bits using the expand function. The expand function is again just a rearrangement of bits; input bits are replicated in a special pattern to expand the 32-bit input to the 48-bit output.

```
function expand(data : vec32) return vec48 is
  --moods inline
begin
  return
    data(32) & data(1) & data(2) &
    ...
    data(31) & data(32) & data(1);
end;
```

This expanded word is then exclusive-ORed with the subkey and fed into a substitute block. This substitutes a different 4-bit pattern for each 6-bit slice of the input pattern (remember that the original input has been expanded from 32 bits to 48 bits, so there are eight substitutions in all). The substitution also has the effect of reducing the output back to 32 bits again. The substitute algorithm first splits the input 48 bits into eight 6-bit slices. Each slice is then used to lookup a substitution pattern for that 6-bit input. This structure is known as the S-block. In the initial implementation, a single ROM is used to store all the substitution patterns. The substitution combines a block index with the input data to form an address, which is then used to lookup the substitution value in the S-block ROM. This address calculation is encapsulated in the smap function.

```
function smap(index : vec3; data : vec6) return vec4 is
  --moods inline
  type S_block_type is
    array(0 to 511) of natural range 0 to 15;
  constant S_block : S_block_type :=
    --moods ROM
  (
  14, 4, 13, 1, 2, 15, 11, 8, 3, 10, 6, 12, 5, 9, 0, 7,
  ...
  2, 1, 14, 7, 4, 10, 8, 13, 15, 12, 9, 0, 3, 5, 6, 11
  );
begin
  return
    vec4(to_unsigned(S_block(to_integer(unsigned(
index & data(1) & data(6) & data(2 to 5)))), 4));
end;
```

The eight substitutions are carried out by the eight calls to smap in the substitute function.

```
function substitute(data : vec48) return vec32 is
  --moods inline
begin
  return
    smap("000",data(1 to 6)) &
    ...
    smap("111",data(43 to 48));
end;
```

The final stage of the datapath transform is the permute function, which is another bit-swapping routine:

```
function permute (data : in vec32) return vec32 is
  --moods inline
begin
  return
    data(16) & data(7) & data(20) & data(21) &
    ...
    data(22) & data(11) & data(4) & data(25);
end;
```

These functions define the whole of the datapath part of the algorithm.

19.4.4 Key Transformations

The encryption key also needs to be transformed a number of times—specifically, before each data transformation, the key is rotated and then a smaller subkey is extracted by selecting 48 of the 56 bits of the key. The rotation is the most complicated part of the key transformation. The 56-bit key is split into two halves and each half rotated by 0, 1, or 2 bits depending on which round of the DES algorithm is being implemented. The direction of the rotation is to the left during encryption and to the right during decryption. The algorithm is split into two functions: do_rotate which, as the name suggests, does the rotation, and key_rotate which calls do_rotate twice, once for each half of the key. The do_rotate function uses a ROM to store the rotate distances for each round, numbered from 0 to 15:

```
function do_rotate
  --moods inline
  (key : in vec28;
  round : natural range 0 to 15;
  encrypt : std_logic)
return vec28 is
  type distance_type is
  array (natural range 0 to 15) of integer range 0 to 2;
  constant encrypt_shift_distance : distance_type :=
  --moods ROM
  (1, 1, 2, 2, 2, 2, 2, 2, 1, 2, 2, 2, 2, 2, 2, 1);
  constant decrypt_shift_distance : distance_type :=
  --moods ROM
  (0, 1, 2, 2, 2, 2, 2, 2, 1, 2, 2, 2, 2, 2, 2, 1);
  variable result : vec28;
begin
  if encrypt = 1 then
  result :=
  vec28(unsigned(key) rol
  encrypt_shift_distance(round));
  else
  result :=
  vec28(unsigned(key) ror
  decrypt_shift_distance(round));
  end if;
  return result;
end;
```

The key_rotate function simply calls the previous function twice:

```
function key_rotate
  --moods inline
  (key : in vec56;
  round : natural range 0 to 15;
  encrypt : std_logic)
  return vec56 is
begin
  return do_rotate(key(1 to 28),round,encrypt) &
  do_rotate(key(29 to 56),round,encrypt);
end;
```

Finally, the key compression function key_compress selects 48 of the 56 bits to pass to the
S-block algorithm.

```
function key_compress(key : in vec56) return vec48 is
  --moods inline
begin
return
  key(14) & key(17) & key(11) & key(24) &
  ...
  key(50) & key(36) & key(29) & key(32);
end;
```

19.5 Initial Synthesis

The design was synthesized by MOODS with delay prioritized first and area prioritized
second. The target technology was the Xilinx Virtex library. Figure 19.2 shows the control
state machine of the synthesized design. The whole state sequence represents the process,
which is a loop as shown by the state transition from the last state (c11) back to the
first (c1).

The first two states c1 and c2 implement the input handshake on signal go to trigger the
process. The DES core is implemented by the remaining states, namely states c3 to c11, which

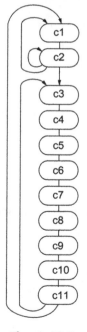

Figure 19.2
Control state machine for initial synthesis.

are in the main loop as shown by the state transition back from c11 to c3, so are executed 16 times. There are nine states in this inner loop, giving a total algorithm length of 146 cycles, including the two states required for the input handshake and 144 for the DES core itself. However, an inspection of the original structure shown in Figure 19.1 suggests that a reasonable target for the inner loop is 2 cycles per round with an optimistic target of 1 cycle. Clearly there is a problem with this design. The synthesis software predicts that this design has the area and delay characteristics shown in Table 19.1 in the line labeled (1).

19.6 Optimizing the Datapath

Examining the nine control states in the main loop and relating these to the mapping of the control graph to the dataflow graph showed that the last 8 cycles were performing the S-block and the first 2 cycles were mainly related to transforming the key. The second state is an overlap state where both key and data transforms are taking place. The problem with the last 8 cycles was fairly self-evident since there are eight substitutions and there are eight control states to perform them. Clearly there was something causing each substitution to be locked into a separate control state and therefore preventing optimization with respect to latency. It wasn't difficult to see what each of these states contained: just register assignments, concatenations and a ROM read operation. It is the last of these that is the problem; the ROM implementation being targeted is a synchronous circuit, so the S-block ROM can only be accessed once per clock cycle—in other words once per control state. It is this that is preventing the datapath operations from being performed in parallel. Attacking this problem is beyond the capabilities of behavioral synthesis because it requires knowledge of the dataflow at a much higher level than can be automatically extracted. The solution therefore requires modification of the original design.

There are two obvious solutions to this problem: either split the S-block into eight smaller ROMs that can therefore be accessed in parallel or make the S-block a non-ROM so that the array gets expanded into a decoder block once for each access, giving eight decoders. The latter solution appears simplest, but it will result in eight 512-way decoders, which will be a very large implementation. The solution of splitting the ROMs is more likely to yield a useful solution. The substitute function was rewritten to have eight mini-ROMs:

```
function substitute(data : vec48) return vec32 is
  --moods inline
  type S_block_type is
    array(0 to 63) of natural range 0 to 15;
  constant S_block0 : S_block_type := ( ... );
  --moods ROM
  ...
```

```
   constant S_block7 : S_block_type := ( ... );
   --moods ROM
begin
   return std_logic_vector(to_unsigned(S_block0(to_integer(
unsigned(data(1) & data(6) & data(2 to 5)))),4)) &
...
std_logic_vector(to_unsigned(S_block7(to_integer(
unsigned(data(43) & data(48) & data(44 to 47)))),4));
end;
```

This was resynthesized and resulted in the control graph shown in Figure 19.3. The inner loop was found to have been reduced to two states, and examination of the last state confirmed that all of the S-block substitutions were being carried out in the one state c4. The key transformations were still split across the two inner states c3 and c4.

One interesting side-effect of this optimization is that it is also a smaller design. MOODS predicts that this design has the area and delay characteristics shown in Table 19.1 in the line labeled (2).

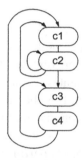

Figure 19.3
Control state machine for optimized S-blocks.

19.6.1 Optimizing the Key Transformations

Examination of the two control states in the main loop, which both contain key transformations, showed that both of these states were performing ROM access and rotate operations. Examination of the original key_rotate function showed that the shift distance ROMs are accessed twice per call, so this turned out to be exactly the same problem as with the S-block ROM. Since ROMs are synchronous, they can only be accessed once per cycle and this forces at least two cycles to be used for the rotate. To solve this, the function can be rewritten to only access the ROMs once per call:

```
if encrypt = 1 then
   distance := encrypt_shift_distance(round);
   result :=
```

```
    vec28(unsigned(key(1 to 28)) rol distance) &
    vec28(unsigned(key(29 to 56)) rol distance);
  else
    distance := decrypt_shift_distance(round);
    result :=
    vec28(unsigned(key(1 to 28)) ror distance) &
    vec28(unsigned(key(29 to 56)) ror distance);
  end if;
```

This was resynthesized and resulted in the control graph shown in Figure 19.4. The inner loop was found to have been reduced to one state (c3) containing both the key and data transformations, which are repeated 16 times. As before, states c1 and c2 implement the input handshake.

So, this optimization means that the target of 1 clock cycle per round of the core was achieved. MOODS predicts that this design has the area and delay characteristics shown in Table 19.1 in the line labeled (3).

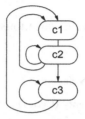

Figure 19.4
Control state machine for optimized key rotate.

19.7 Final Optimization

It was recognized that the key_rotate function could be simplified by rethinking the rotate algorithm such that a right rotate of 1 bit was replaced by a left rotate of 27 bits (for a 28-bit word). This eliminates a conditional statement, which it was felt could be preventing some optimizations from taking place. This means that there was no need to have a different algorithm for encryption and decryption. This led to the following rework:

```
function key_rotate
  --moods inline
  (key : vec56;
  round : natural range 0 to 15;
  encrypt : std_logic)
return vec56 is
  type distance_type is
  array (natural range 0 to 31) of integer range
  0 to 31;
  constant shift_distance : distance_type :=
    --moods ROM
```

```
    ( 0, 1, 2, 2, 2, 2, 2, 2,
      1, 2, 2, 2, 2, 2, 2, 1,
     27, 27, 26, 26, 26, 26, 26, 26,
     27, 26, 26, 26, 26, 26, 26, 27);
  variable distance : natural range 0 to 31;
begin
  distance := shift_distance(to_integer(unsigned(
  encrypt & to_unsigned(round,4))));
  return vec28(unsigned(key(1 to 28)) ror distance) &
  vec28(unsigned(key(29 to 56)) ror distance);
end;
```

The state machine for this design was basically the same as for the previous design as shown in Figure 19.4. It was found that this version was slightly slower than the previous design but significantly smaller. MOODS predicts that this design has the area and delay characteristics shown in Table 19.1 in the line labeled (4).

19.8 Results

The results predicted by MOODS for all the variations of the design discussed so far are summarized in the following table:

Table 19.1 Physical metrics for single DES designs

Design	Area (Slices)	Latency (Cycles)	Clock (ns)	Throughput (MB/s)
(1) Initial design	552	146	7.8	7.12
(2) Optimized S blocks	426	34	7.1	35.2
(3) Optimized key	489	18	7.1	62.6
(4) Optimized branch	307	18	8.4	52.9

It can be seen that design (3) is the fastest, but design (4) is the smallest. Figure 19.5 plots area vs. throughput for these four designs. The X-axis represents the area of the design and the Y-axis the throughput.

19.9 Triple DES

19.9.1 Introduction

Building on this, the DES core developed previously was used as the core for a Triple-DES implementation. The idea of triple DES is that data is encrypted three times. The rationale for choosing three iterations and the advantages and disadvantages of this are explained in [5]. A common form of Triple DES is known as EDE2, which means data is encrypted, decrypted and then encrypted again using two different keys. The first key is used for both encryptions and the second key for the decryption. There are obviously a number of different trade-offs

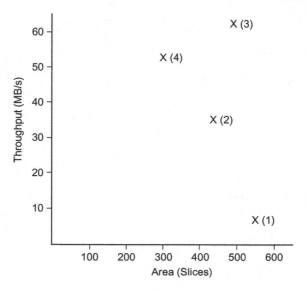

Figure 19.5
Area vs. throughput for all DES designs.

that can be made in this design. Each of these is examined in the following sections. In all cases, the smallest implementation (design (4)) was used as the DES core.

19.9.2 Minimum Area Iterative

To achieve a minimum area implementation, a single DES core is used for all three stages. The data is passed through this core three times with the different permutations of keys and encryption mode to achieve the EDE2 algorithm. Two different styles of VHDL were tried. These differed in the method used to select the different inputs for each encryption step. The first style used a case statement and the second style used indexed arrays. The case statement style results in the following VHDL design:

```
library ieee;
use ieee.std_logic_1164.all;
entity tdes_ede2_iterative is
  port(
    plaintext : in std_logic_vector(1 to 64);
    key1 : in std_logic_vector(1 to 64);
    key2 : in std_logic_vector(1 to 64);
    encrypt : in std_logic;
    go : in std_logic;
    ciphertext : out std_logic_vector(1 to 64);
    done : out std_logic);
end;
architecture behavior of tdes_ede2_iterative is
```

```
    ...
  begin
   process
     variable data : vec64;
     variable key : vec56;
     variable mode : std_logic;
   begin
    wait until go = 1;
    done <= 0;
    wait for 0 ns;
    data := plaintext;
    for i in 0 to 2 loop
      case i is
        when 1 =>
          key := key_reduce(key2);
          mode := not encrypt;
        when others =>
          key := key_reduce(key1);
          mode := encrypt;
      end case;
      data := des_core(data,key,mode);
    end loop;
    ciphertext <= data;
    done <= 1;
   end process;
  end;
```

It can be seen that this uses a case statement to select the appropriate key and encryption mode for each iteration. The characteristics of the case statement solution are shown in Table 19.2 in the line labeled (5). The core DES algorithm accounts for 48 cycles (3 iterations of 16 rounds with 1 cycle per round), leaving an additional overhead of 3 cycles. This additional 3 cycles is due to the case statement selection of the key, which adds an extra cycle per iteration of the core. The second style uses arrays to store the keys and modes and then indexes these arrays to set the key and mode for each iteration. The process becomes:

```
process
  ...
  type keys_type is array (0 to 2) of vec56;
  variable keys : keys_type;
  type modes_type is array (0 to 2) of std_logic;
  variable modes : modes_type;
begin
  ...
  modes := (encrypt, not encrypt, encrypt);
  keys := (key_reduce(key1),
  key_reduce(key2),
  key_reduce(key1));
  for i in 0 to 2 loop
    data := des_core(data,keys(i),modes(i));
  end loop;
  ...
```

It was found that the latency was the same as the case statement solution but the area was approximately 25% larger. This overhead is mostly due to the use of the register arrays, which add up to about 200 extra flip-flops. Clearly the case statement design is the most efficient of the two and so this solution was kept and the array style solution discarded.

19.9.3 Minimum Latency Pipelined

To achieve minimum latency between samples, three DES cores are used to form a pipeline. Data samples can then be fed into the pipeline every 18 cycles (the latency of the single core), although the time taken for a result to be generated is 50 cycles because of the pipeline length. The circuit is simply three copies of the single-DES process:

```
architecture behavior of tdes_ede2_pipe is
  ...
  signal intermediate1, intermediate2 : vec64;
begin
  process
  begin
    wait until go = 1;
    intermediate1 <=
    des_core(plaintext,key_reduce(key1),encrypt);
  end process;
  process
  begin
    wait until go = 1;
    intermediate2 <=
    des_core(intermediate1,key_reduce(key2),not
    encrypt);
  end process;
  process
  begin
    wait until go = 1;
    done <= 0;
    wait for 0 ns;
    ciphertext <=
    des_core(intermediate2,key_reduce(key1),
    encrypt);
    done <= 1;
  end process;
end;
```

Note how the done output is driven only by one of the cores; âŁ" this will give the right result provided all three cores synthesize to the same delay, which in practice they will. This design decision alleviates the need to have handshaking between the cores. MOODS predicts that this design has the area and delay characteristics shown in Table 19.2 in the line labeled (6). The state machine in Figure 19.6 shows the three independent processes. For example, the first

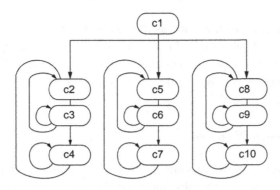

Figure 19.6
Control state machine for pipelined triple-DES.

process is represented by states c2, c3, and c4. The first two states perform the handshaking on go and c4 implements the DES core with its 16 iterations. State c7 is the second DES core and c10 the third.

19.10 Comparing the Approaches

The physical metrics of the previous section are the predicted values given by MOODS. To get a more accurate assessment of the design, RTL synthesis of the structural VHDL output of MOODS is required. This was carried out using Mentor Graphics Leonardo Spectrum RTL synthesis suite. These results can be finessed further by carrying out placement and routing using the Xilinx Integrated Software Environment (ISE) Foundation suite. The results predicted by all three tools (MOODS, Leonardo and Foundation) for the three approaches (DES, Iterative TDES and Pipelined TDES) are shown in Table 19.2. In all cases, the design was optimized during RTL synthesis using the vendors' default optimization settings—a combination of area and delay optimization—with maximum optimization effort. Placement

Table 19.2 Predicted Results for MOODS, Leonard and Foundation Tools.

Design	Tool	Area (slices)	Latency (cycles)	Clock (ns)	Throughput (MB/s)
(4)	MOODS	307	18	8.4	52.9
DES	Leonardo	258		13.4	33.2
	Foundation	274		18.4	24.2
(5)	MOODS	500	53	8.4	18.0
Iterative	Leonardo	381		13.7	11.0
TDES	Foundation	422		17.8	8.5
(6)	MOODS	920	18	8.4	52.9
Pipelined	Leonardo	774		13.7	32.4
TDES	Foundation	826		18.4	24.2

and routing was performed with an unreachable clock period to force Foundation to produce the fastest design.

This shows that MOODS tends to overestimate the area of the design and underestimate the delay. Both of these are expected outcomes. The tendency to overestimate area is because it isn't possible to predict the effect of logic minimization when working at the behavioral level. The tendency to underestimate delay is because it isn't possible to predict routing delays.

19.11 Summary

This chapter has shown that it is possible to design complex algorithms such as DES using the abstraction of high-level VHDL and get a synthesizable design. However, the synthesis process is not and cannot ever be fully automated—human guidance is still necessary to optimize the design's structure to get the best from the synthesis tools. Nevertheless, the modifications are high-level design decisions and the final design is still readable and abstract. There has been no need to descend to low-level VHDL to implement DES. The implementations of Triple-DES show how VHDL code can easily be reused when written at this level of abstraction. It is quite an achievement to implement the DES and two implementations of the Triple-DES algorithm in four working days, including testing, and this demonstrates the kind of productivity that results from the application of behavioral synthesis tools.

For more details of the analysis of these techniques, the reader is referred to the technical papers [1–3].

References

[1] A.D. Brown, D. Milton, A. Rushton, P.R. Wilson, Behavioural synthesis utilising recursive definitions, IET Comput. Digit. Tech. 6 (6) (2012) 362–369.
[2] M. Sacker, A.D. Brown, A.J. Rushton, P.R. Wilson, A behavioral synthesis system for asynchronous circuits, IEEE Trans. Very Large Scale Integr. Syst. 12 (9) (2004) 978–994.
[3] P.R. Wilson, A.D. Brown, DES in 4 days using behavioural modeling and simulation, in: IEEE International Behavioral Modeling and Simulation Conference, BMAS 2005, San Jose, USA, 2005.

PART 5

Fundamental Techniques

This final part of the book provides an insight into the basic building blocks used in everyday digital designs implemented on FPGAs. These include such low-level functions as latches, registers, counters, logic functions, serialization and de-serialization, finite state machines, decoders, multiplexers and digital arithmetic.

CHAPTER 20

Latches, Flip-Flops, and Registers

20.1 Introduction

There are different types of storage elements that will occur from different HDL code, and it is important to understand each of them, so that the correct one results when a design is synthesized. Often bugs in hardware happen due to misunderstandings about what effect a particular HDL construct (in VHDL or Verilog) will have on the resulting synthesized hardware. In this chapter, we will introduce the three main types of storage elements that can be synthesized from VHDL or Verilog to an FPGA platform: which are latches, flip-flops, and registers.

20.2 Latches

A latch can be simply defined as a level-sensitive memory device. In other words, the output depends purely on the value of the inputs. There are several different types of latch, the most common being the D latch and the SR latch. First consider a simple D latch as shown in Figure 20.1. In this type of latch, the output (Q) follows the input (D), but only when the Enable (En) is high. The full definition is in fact a level-sensitive D latch, and the assumption made in this book is that whenever we refer to a latch, it is always level sensitive. It is worth noting that latches are not particularly useful in FPGA design as they are obviously asynchronous and therefore can cause timing issues. In practice it is much better to use synchronous D-types, as will be introduced later in this chapter; however, it is worth looking at basic latches to see the differences.

The notation on the Enable signal (C1) and the Data input (1D) denotes that they are linked together. Also notice that the output Q is purely dependent on the level of D and the Enable. In other words, when the Enable is high, then Q = D. So, as previously stated, this is a level-sensitive latch.

The VHDL that represents this kind of level-sensitive D latch is shown here:

```
library ieee;
use ieee.std_logic_1164.all;
entity latch is
  port (
    d : in std_logic;
```

Design Recipes for FPGAs. http://dx.doi.org/10.1016/B978-0-08-097129-2.00020-9

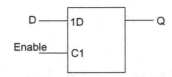

Figure 20.1
Basic D latch symbol.

```
    en : in std_logic;
    q : out std_logic
    );
end entity latch;

architecture beh of latch is
begin
process (d, en) is
begin
  if (en = '1') then
    q <= d;
  end if;
end process;
end architecture beh;
```

We can implement a very similar model using Verilog with the code given following this paragraph. In this case the Verilog uses the *always* statement to check for any changes on the enable or the d inputs and if enable (*en*) is high, then the output will also change.

```
module dlatch (
  d,  // Data Input
  en, // Enable Input
  q   // Latch Output
);

input d;
input en;

output q;
reg q;

always @ ( en or d )
if (en) begin
  q <= d;
end

endmodule
```

This is an example of an *incomplete if* statement, where the condition *if* (*en* = 1) is given, but the *else* condition is not defined. Both *d* and *en* are in the sensitivity list and so this could be combinatorial, but due to the incomplete definition of *en*, then an implied latch occurs, that is, storage. This aspect of storage is important when we are developing models, particularly behavioral as in this case (i.e., the structure is not explicitly defined), as we may end up with

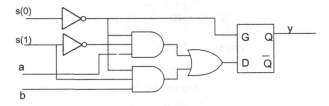

Figure 20.2
Synthesized latch.

latches in our design even though we think that we have created a model for a purely combinatorial circuit.

Other instances when this may occur are the incomplete definition of case statements. For example, consider this simple VHDL example:

```
case s is
  when "00" => y <= a;
  when "10" => y <= b;
  when others => null;
end case;
```

In this statement, it is incomplete and so instead of a simple combinatorial circuit, a latch is therefore implied. The resulting synthesized circuit is shown in Figure 20.2.

Similar outcomes would happen if a Verilog case statement was incomplete:

```
case s is
  2'b00 : y=a;
  2'b10" : y=b;
end case;
```

In both cases it is important to ensure that all the states are defined to avoid this happening, and in Verilog this is done by using the *default* option in the case statement. Therefore, in our simple example, to ensure that all the possible values of *s* are covered, the default line will specify an output in those cases as shown here:

```
case s is
  2'b00 : y=a;
  2'b10" : y=b;
  default : y=0;
end case;
```

20.3 Flip-Flops

In contrast to the level-triggered latch, the flip-flop changes state when an edge occurs on an enable or a clock signal. This is the cornerstone of synchronous design, with an important building block being the D-type flip-flop, as shown in Figure 20.3. The output (Q) will take on

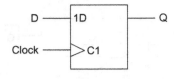

Figure 20.3
D type flip-flop.

the value of the input (D) on the rising edge of the clock signal. The triangle on the symbol denotes a clock signal and, in the absence of a circle (notation for active low), the definition is for a rising edge to activate the flip-flop.

The equivalent VHDL code is of the form shown as follows:

```
library ieee;
use ieee.std_logic_1164.all;
entity dff is
  port (
    d : in std_logic;
    clk : in std_logic;
    q : out std_logic
  );
end entity dff;

architecture simple of dff is
begin
process (clk) is
begin
  if rising_edge(clk) then
    q <= d;
  end if;
end process;
end architecture simple;
```

Notice that, in this case, d does not appear in the sensitivity list, as it is not required. The flip-flop will only do something when a rising edge occurs on the clock signal (clk). There are a number of different methods of describing this functionality, all of them equivalent. In this case, we have explicitly defined the clk signal in the sensitivity list.

An alternative method in VHDL would be to have no sensitivity list, but to add a wait on statement inside the process. The equivalent architecture would be as follows:

```
architecture wait_clk of dff is
begin
process is
begin
  if rising_edge(clk) then
    q <= d;
  end if;
  wait on clk;
end process;
end architecture simple;
```

This could also be defined using a slightly different implementation of the wait statement, to wait for an event on clk and then check if it was high:

```
architecture wait_clk of dff is
begin
process is
begin
  wait until clk'event and clk = '1';
  q <= d;
end process;
end architecture simple;
```

We have also perhaps used a more complex definition of the rising_edge function than is required (or may be available in all simulators or synthesis tools). The alternative simple method is to use the clock in the sensitivity list and then check that the value of clock is 1 for rising edge or 0 for falling edge. The equivalent VHDL for a rising edge D-type flip-flop is given. Notice that we have used the implicit sensitivity list (using a wait on clk statement) as opposed to the explicit sensitivity list, although we could use either interchangeably.

```
architecture rising_edge_clk of dff is
begin
process is
begin
  if (clk = 1) then
    q <= d;
  end if;
  wait on clk;
end process;
end architecture simple;
```

We can also implement this type of flip-flop using Verilog as shown here:

```
module dff (
  d,   // Data Input
  clk, // Clock Input
  q    // Latch Output
);

input d;
input clk;

output q;
reg q;

always @ ( posedge clk )
begin
  q <= d;
end

endmodule
```

We can extend this basic model for a D-type to include an asynchronous set and reset function. If they are asynchronous, this means that they could happen whether there is a clock

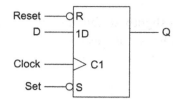

Figure 20.4
D-type flip-flop with asynchronous set and reset.

edge or not; therefore they need to be added to the sensitivity list of the model. The symbol for such a flip-flop, assuming active low set and reset, would be as shown in Figure 20.4.

The VHDL is extended from the simple dff model previously given to include the asynchronous set and reset as shown in the following:

```
library ieee;
use ieee.std_logic_1164.all;
entity dff_sr is
  port (
    d : in std_logic;
    clk : in std_logic;
    nrst : std_logic;
    nset : in std_logic;
    q : out std_logic
  );
end entity dff_sr;

architecture simple of dff_sr is
begin
process (clk, nrst, nset) is
  begin
    if (nrst = '0') then
      q <= '0';
    elsif (nset = 1) then
      q <= '1';
    elsif rising_edge(clk) then
      q <= d;
    end if;
  end process;
end architecture beh;
```

As for the basic D type flip-flops, we could use a variation of the check for the clock edge, although due to the fact that we have three possible input state control variables (set, reset, and clk), it is not enough now to check whether the clock is high (for a rising edge flip-flop). It is necessary to check that the clock is high and that an event has occurred.

Notice that this model may cause interesting behavior when synthesized, as the reset will always be checked before the set and so there is a specific functionality that allows the concurrent setting of the set and reset variables, but the reset will take precedence.

We can also implement this behavior in a very similar fashion in Verilog as shown in the following listing:

```
module dff_asr (
  d,  // Data Input
  s,  // Set Input
  r,  // Reset Input
  q   // Latch Output
);

input d;
input s;
input r;

output q;
reg q;

always @ ( s or r or clk )
if (~r) begin
  q <= 1b'0;
end elseif (s) begin
  q <= 1b'1;
else begin
  q <= d;
end

endmodule
```

Finally, when considering the use of transitions between 0 and 1, there are issues with synthesis and simulation when using the different approaches. For example, with the standard logic package (std_logic variables), the transitions are strictly defined and so we may have the case of high impedance or "don't care" states occurring during a transition. This is where the rising_edge function and its opposite, the falling_edge function, are useful as they simplify all these options into a single function that handles all the possible transition states cleanly.

It is generally best, therefore, to use the rising_edge or falling_edge functions in VHDL (and posed and negedge in Verilog) wherever possible to ensure consistent and interoperable functionality of models.

It is also worth considering a synchronous set or reset function, so that the clock will be the only edge that is considered. The only caveat with this type of approach is that the set and reset signals should be checked immediately following the clock edge to make sure that concurrent edges on the set or reset signals have not occurred.

20.4 Registers

Registers use a bank of flip-flops to load and store data in a bus. The difference between a basic flip-flop and a register is that, while there is a data input, clock and usually a reset (or clear), there is also a *load* signal that defines whether the data on the input is to be loaded onto the register or not. The VHDL code for an example 8-bit register would be as follows:

```vhdl
library ieee;
use ieee.std_logic_1164.all;
entity register is
  generic (
    n : natural := 8
  );
  port (
    d : in std_logic_vector(n-1 downto 1);
    clk : in std_logic;
    nrst : in std_logic;
    load : in std_logic;
    q : out std_logic_vector(n-1 downto 1)
  );
end entity register;

architecture beh of register is
begin
process (clk, nrst) is
  begin
    if (nrst = '0') then
      q <= (others => '0');
    elsif (rising_edge(Clock) and (load = 1)) then
      q <= d;
    end if;
  end process;
end architecture beh;
```

This can also use a more indented form of *if-then-else* structure to separate the check on the clk and the load values as shown here:

```vhdl
library ieee;
use ieee.std_logic_1164.all;
entity register is
  generic (
    n : natural := 8
  );
  port (
    d : in std_logic_vector(n-1 downto 1);
    clk : in std_logic;
    nrst : in std_logic;
    load : in std_logic;
    q : out std_logic_vector(n-1 downto 1)
  );
end entity register;

architecture beh of register is
begin
process (clk, nrst) is
  begin
    if (nrst = '0') then
      q <= (others => '0');
    elsif (rising_edge(Clock) then
```

```
        if (load = '1')) then
              q <= d;
         end if;
    end if;
  end process;
end architecture beh;
```

Notice that although there are four inputs (clk, nrst, load, and d), only clk and nrst are included in the process sensitivity list. If load and d change, then the process will ignore these changes until the clk rising edge or nrst goes low. If the load is not used, then the register will load the data on every clock rising edge unless the reset is low. This can be useful in applications such as pipelining, where efficiency is paramount. The VHDL for this slightly simpler register is given here:

```
library ieee;
use ieee.std_logic_1164.all;
entity reg_rst is
  port (
  d,
  clk,
  nrst : in std_logic;
  q : out std_logic
  );
end entity reg_rst;

architecture beh of reg_rst is
begin
process (clk, nrst) is
  begin
    if (nrst = '0') then
      q <= '0';
    elsif rising_edge(clk) then
      q <= d;
    end if;
  end process;
end architecture beh;
```

In a similar manner we can write a register model with an input d, load, and nrst control signals, with a clock input and the output q. In this case we use the posedge (positive edge) of the clock and the nest variable as the sensitivity list to the *always* block.

```
module register (
  d,
  clk,
  nrst,
  load,
  q
  );

parameter n =8;

input [n-1:0] d;
input clk;
input nrst;
```

```
input load;

output q;
reg [n-1:0] q;

always @ (posedge clk or nrst)

if(nrst == 1b'0) then begin
  q <= 0;
end
else
  if (load == 1b'1) then begin
    q <= d;
  end
end

endmodule
```

20.5 Summary

In this chapter, the basic type of latch and register have been introduced and examples given. This is a fundamental building block of synchronous digital systems and is the basis of RTL (Register Transfer Logic) design with VHDL or Verilog.

CHAPTER 21

ALU Functions

21.1 Introduction

A central part of microprocessors is the ALU (Arithmetic Logic Unit). This block in a processor takes a number of inputs from registers and as its name suggests carries out either logic functions (such as NOT, AND, OR, and XOR) on the inputs, or arithmetic functions (addition or subtraction at a minimum), although it must be noted that these will be integer (or fixed point, potentially) and not floating point. This chapter of the book will describe how these types of low-level logic and arithmetic functions can be implemented in VHDL and Verilog. In this chapter, in some cases we have used bit and in other std_logic, both of which are valid. In practice, most designers will use the more complete definition in std_logic (and the vector extension std_logic_vector); however, both may be used. In the case of std_logic, the IEEE library must be included.

21.2 Logic Functions in VHDL

If we consider a simple inverter in VHDL, we can develop a single inverter which takes a single input bit, inverts it and applies this to the output bit. This simple VHDL is shown as follows:

```
library ieee;
use ieee.std_logic_1164.all;
entity inverter is
  port (
    a : in std_logic;
    q : out std_logic
  );
end entity inverter;
architecture simple of inverter is
begin
  q <= not a;
end architecture simple;
```

Clearly the inputs and output are defined as single std_logic pins, with direction in and out, respectively. The logic equation is also intuitive and straightforward to implement. We can extend this to be applicable to n bit logic busses by changing the entity (the architecture

Design Recipes for FPGAs. http://dx.doi.org/10.1016/B978-0-08-097129-2.00021-0
295

remains the same) and simply assigning the input and outputs the type std_logic_vector
instead of std_logic, as follows:

```
library ieee;
use ieee.std_logic_1164.all;
entity bus_inverter is
  port (
    a : in std_logic_vector(15 downto 0);
    q : out std_logic_vector(15 downto 0)
  );
end entity bus_inverter;
architecture simple of bus_inverter is
begin
  q <= not a;
end architecture simple;
```

As can be seen from the VHDL, we have defined a specific 16-bit bus in this example, and
while this is generally fine for processor design with a fixed architecture, sometimes it is
useful to have a more general case, with a configurable bus width. In this case we can modify
the entity again to make the bus width a parameter of the model, which highlights the power
of using generic parameters in VHDL.

```
library ieee;
use ieee.std_logic_1164.all;
entity n_inverter is
  generic (
    n : natural := 16
  );
  port (
    a : in std_logic_vector((n−1) downto 0);
    q : out std_logic_vector((n−1) downto 0)
      );
end entity n_inverter;
architecture simple of n_inverter is
begin
  q <= not a;
end architecture simple;
```

We can of course create separate models of this form to implement multiple logic functions,
but we can also create a compact multiple function logic block by using a set of configuration
pins to define which function is required. If we define a general logic block that has 2 n-bit
inputs (A and B), a control bus (S) and an n-bit output (Q), then by setting the 2-bit control
word (S) we can select an appropriate logic function according to the following table:

S	Function
00	Q <= NOTA
01	Q <= AANDB
10	Q <= AORB
11	Q <= AXORB

Clearly we could define more functions, and this would require more bits for the select function (S) which could also be defined using a generic, but this limited set of functions demonstrates the principle involved. We can define a modified entity as shown:

```
library ieee;
use ieee.std_logic_1164.all;
entity alu_logic is
      generic (
            n : natural := 16
      );
      port (
            a : in std_logic_vector((n-1) downto 0);
            b : in std_logic_vector((n-1) downto 0);
            s : in std_logic_vector(1 downto 0);
            q : out std_logic_vector((n-1) downto 0)
      );
end entity alu_logic;
```

Now, depending on the value of the input word (S), the appropriate logic function can be selected. We can use the case statement introduced in Chapter 3, A VHDL Primer to define each state of S and which function will be carried out in a very compact form of VHDL:

```
architecture basic of alu_logic is
begin
      case s is
            when "00" => q <= not a;
            when "01" => q <= a and b;
            when "10" => q <= a or b;
            when "11" => q <= a xor b;
      end case;
end architecture basic;
```

Clearly this is an efficient and compact method of defining the combinatorial logic for each state of the control word (S), but great care must be taken to assign values for every combination to avoid inadvertent latches being introduced into the logic when synthesized. To avoid this, a synchronous equivalent could also be implemented that only applied the logic function on the clock edge specified.

In this example, all of the possible combinations are specified; however, in order to avoid possible inadvertent latches being introduced, it would be good practice to use a "when others" statement to cover all the unused cases.

21.2.1 1-bit Adder

The arithmetic *heart* of an ALU is the addition function (Adder). This starts from a simple 1-bit adder and is then extended to multiple bits, to whatever size addition function is required in the ALU. The basic design of a 1-bit adder is to take two logic inputs (a and b) and produce a sum and carry output according to the following truth table:

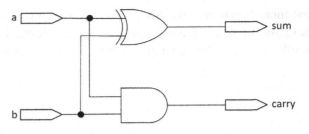

Figure 21.1
1-bit adder.

a	b	sum	carry
0	0	0	0
0	1	1	0
1	0	1	0
1	1	1	1

This can be implemented using simple logic with a 2 input AND gate for the carry, and a 2 input XOR gate for the sum function, as shown in Figure 21.1.

This function has a carry out (carry), but no carry in, so to extend this to multiple bit addition, we need to implement a carry in function (cin) and a carry out (cout) as follows:

a	b	cin	sum	Cout
0	0	0	0	0
0	1	0	1	0
1	0	0	1	0
1	1	0	0	1
0	0	1	1	0
0	1	1	0	1
1	0	1	0	1
1	1	1	1	1

With an equivalent logic function as shown in Figure 21.2:

This can be implemented using standard VHDL logic functions with bit inputs and outputs as follows. First, define the entity with the input and output ports defined using bit types:

```
entity full_adder is
  port (sum, co : out bit;
        a, b, ci : in bit);
end entity full_adder;
```

Then the architecture can use the standard built-in logic functions in a dataflow type of model, where logic equations are used to define the behavior, without any delays implemented in the model.

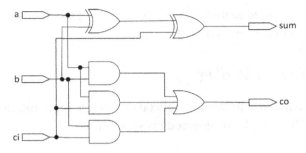

Figure 21.2
1-bit adder with carry-in and carry-out.

```
architecture dataflow of full_adder is
begin
  sum <= a xor b xor ci;
  co <= (a and b) or
        (a and ci) or
        (b and ci);
end architecture dataflow;
```

This model is now a simple building block that we can use to create multiple bit adders structurally by linking a number of these models together.

21.3 Structural n-Bit Addition

Using the simple 1-bit full adder defined previously, it is a simple matter to create a multiple bit full adder using this model as a building block. As an example, to create a 4-bit adder, with a single bit carry in and single bit carry out, we can define a VHDL model as shown here:

```
entity four_bit_adder is
  port (sum: out bit_vector (3 downto 0); co : out bit;
        a, b : in bit_vector (3 downto 0); ci : in bit);
end entity four_bit_adder;

architecture simple of four_bit_adder is
  signal carry : bit_vector (3 downto 1);
begin
  fa0 : entity work.full_adder
port map (sum(0),carry(1),a(0),b(0),ci);
  fa1 : entity work.full_adder
port map (sum(1),carry(2),a(1),b(1),carry(1));
  fa2 : entity work.full_adder
port map (sum(2),carry(3),a(2),b(2),carry(2));
  fa3 : entity work.full_adder
port map (sum(3),co,a(3),b(3),carry(3));
end architecture simple;
```

This can obviously be extended to multiple bits by repeating the component use in the architecture for as many bits as are required.

21.4 Logic Functions in Verilog

If we consider a simple inverter in Verilog, this takes a single input bit, inverts it and applies this to the output bit. This simple Verilog code is shown here:

```
module inverter (
  q,
  a
  );

input a;
output q;

assign q = ~ a;

endmodule
```

Clearly the inputs and output are defined as single std_logic pins, with direction in and out respectively. The logic equation is also intuitive and straightforward to implement. We can extend this to be applicable to n bit logic busses by changing the inputs and outputs (the architecture remains the same) into bus types as follows:

```
module bus_inverter (
  q,
  a
  );

input [15:0] a;
output [15:0] q;

assign q = ~ a;

endmodule
```

As can be seen from the Verilog, we have defined a specific 16-bit bus in this example, and while this is generally fine for processor design with a fixed architecture, sometimes it is useful to have a more general case, with a configurable bus width. In this case we can modify the Verilog again to make the bus width a parameter of the model:

```
module n_inverter (
  q,
  a
  );

param n = 16;
input [n-1:0] a;
output [n-1:0] q;
```

```
assign q = ~ a;

endmodule
```

We can of course create separate models of this form to implement multiple logic functions, but we can also create a compact multiple function logic block by using a set of configuration pins to define which function is required, as we did in the case for the VHDL. Clearly we could define more functions, and this would require more bits for the select function (S), but this limited set of functions demonstrates the principle involved.

Now, depending on the value of the input word (S), the appropriate logic function can be selected. We can use the case statement introduced in Chapter 3 of this book to define each state of S and which function will be carried out in a very compact form. As in the VHDL case, this is an efficient and compact method of defining the combinatorial logic for each state of the control word (S), but great care must be taken to assign values for every combination to avoid inadvertent latches being introduced into the logic when synthesized.

21.5 Configurable n-Bit Addition

While the structural approach is useful, it is clearly cumbersome and difficult to configure easily. A more sensible approach is to add a generic (parameter) to the model to enable the number of bits to be customized. For example, if we define an entity to add two logic vectors (as opposed to bit vectors used previously), the entity will look something like this:

```
library IEEE;
use IEEE.std_logic_1164.all;

entity add_beh is
  generic(top : natural := 15);
    port (
      a   : in std_logic_vector (top downto 0);
      b   : in std_logic_vector (top downto 0);
      cin : in std_logic;
      sum : out std_logic_vector (top downto 0);
      cout : out std_logic
    );
end entity add_beh;
```

As can be seen from this entity, we have a new parameter, top, which defines the size of the input vectors (a and b) and the output sum (cout). We can then use the same original logic equations that we defined for the initial 1-bit adder and use more behavioral VHDL to create a much more readable model:

```
architecture behavior of add_beh is
begin
  adder:
```

```
process(a,b,cin)
  variable carry : std_logic;
  variable tempsum : std_logic_vector(top downto 0);
  begin
    carry := cin;
    for i in 0 to top loop
      tempsum(i) := a(i) xor b(i) xor carry;
      carry := (a(i) and b(i)) or (a(i) and carry) or (b(i) and carry);
    end loop;
    sum <= tempsum;
    cout <= carry;
end process adder;
end architecture behavior;
```

This architecture shows how a single process (with sensitivity list a,b,cin) is used to encapsulate the addition. The process is activated when a,b or cin changes. A for loop is used to calculate a temporary sum (tempsum) that increments each time around the loop if required and the final value is assigned to the output sum. Also, a stage by stage carry is calculated and used each time around the loop. After the final loop, the value of carry is used to become the final carry out.

21.6 Two's Complement

An integral part of subtraction in logic systems is the use of *two's complement*. This enables us to execute a subtraction function using only an adder rather than requiring a separate subtraction function. Two's complement is an extension to the basic ones' complement (or basic inversion of bits) previously considered.

If we consider an unsigned number system based on 4 bits, then the range of the numbers is 0 to 15 (0000 to 1111). If we consider a signed system, however, the most significant bit (MSB) is considered to be the sign (+ or −) of the number system and therefore the range of numbers with 4 bits will instead be from −8 to +7. The method of conversion from positive to negative number in binary logic is a simple two-stage process of first inverting all the bits and then adding 1 to the result.

Consider an example. Take a number 00112. In signed number form, the MSB is 0, so the number is positive and the lower three bits 011 can be directly translated into decimal 3. To get the two's complement (−3), we first invert all the bits to get 1100, and then add a single bit to get the final two's complement value 1101. To check that this is indeed the inverse in binary, simple add the number 0011 to its two's complement 1101 and the result should be 0000.

In a signed system the range of numbers is $-(2^N - 1)$ to $+(2^N - 1 - 1)$ whereas in the unsigned system the range is defined by 0 to $+(2^N - 1)$. The signed system allows both positive and negative numbers; however, the maximum magnitude is effectively half the magnitude of the unsigned system.

This function can be implemented simply in VHDL using the following model:

```
library ieee;
use ieee.std_logic_1164.all;
use ieee.numeric_std.all;

entity twoscomplement is
  generic (
        n : integer := 8
              );
  port (
        input : in std_logic_vector((n-1) downto 0);
        output : out std_logic_vector((n-1) downto 0)
  );
end;

architecture simple of twoscomplement is
begin
  process(input)
      variable inv : unsigned((n-1) downto 0);
  begin
        inv := unsigned(NOT input);
        inv := inv + 1;
        output <= std_logic_vector(inv);
  end process;
end;
```

As can be seen from the VHDL, we operate using logic functions first (NOT) and then convert to unsigned to utilize the addition function (inv + 1), and finally convert the result back into a std_logic_vector type. Also notice that the generic n allows this model to be configured for any data size. In this example, the test bench is used to check that the function is operating correctly by using two test circuits back to back, inverting and re-inverting the input word and checking that the function returns the same value. While this does not guarantee correct operation (the same bug could be present in both transforms!), it is a simple quick check that is very useful and makes generation of test data and checks very easy, as the input and final output signal check can be XORed to check for differences.

```
library ieee;
use ieee.std_logic_1164.all;
use ieee.numeric_std.all;

entity twoscomplementtest is
end twoscomplementtest ;

architecture stimulus of twoscomplementtest is
      signal rst : std_logic := '0';
      signal clk : std_logic:='0';
      signal count : std_logic_vector (7 downto 0);
      signal inverse : std_logic_vector (7 downto 0);
```

```
        signal check : std_logic_vector (7 downto 0);
        component twoscomplement
            port(
                    input : in std_logic_vector(7 downto 0);
                    output : out std_logic_vector(7 downto 0)
                );
        end component;
        for all : twoscomplement use entity work.twoscomplement ;
begin
    CUT1: twoscomplement port map(input=>count,output=>inverse);
    CUT2: twoscomplement port map(input=>inverse,output=>check);

    -- clock and reset process
    clk <= not clk after 1 us;
    process
    begin
      rst<='0','1' after 2.5 us;
      wait;
    end process;

    -- generate data
    process(clk, rst)
      variable tempcount : unsigned(7 downto 0);
    begin
      if rst = '0' then
        tempcount := (others => '0');
      elsif rising_edge(clk) then
        tempcount := tempcount + 1;
      end if;
      count <= std_logic_vector(tempcount );
    end process;
end;
```

21.7 Summary

This chapter has introduced the key elements required in an Arithmetic and Logic Unit of a processor. Whether the designer needs to implement a complete ALU from scratch, or just to understand the behavior of an existing architecture, these functions are very useful in analyzing the behavior of ALUs and processors.

CHAPTER 22

Finite State Machines in VHDL and Verilog

22.1 Introduction

Finite State Machines (FSMs) are at the heart of most digital design. The basic idea of an FSM is to store a sequence of different unique states and transition between them depending on the values of the inputs and the current state of the machine. The FSM can be of two types: Moore (where the output of the state machine is purely dependent on the state variables) and Mealy (where the output can depend on the current state variable values AND the input values).[1] The general structure of an FSM is shown in Figure 22.1:

22.2 State Transition Diagrams

One method of describing Finite State Machines from a design point of view is using a state transition diagram (bubble chart) which shows the states, outputs and transition conditions.[2] A simple state transition diagram is shown in Figure 22.2.

Interpreting this state transition diagram, it is clear that there are four bubbles (states). The transitions are controlled by two signals (*rst* and *choice*), both of which could be represented by bit or std_logic types (or another similar logic type). There is an implicit clock signal, which we shall call *clk* and the single output *out1*.

22.3 Implementing Finite State Machines in VHDL

This transition diagram can be implemented using a case statement in a process using the following VHDL:

```
library ieee;
use ieee.std_logic_1164.all;

entity fsm is
  port(
```

---

[1] Although in fact it is also possible to have a hybrid of the two approaches.
[2] Another method is the algorithmic state machine approach, which shows actions and decisions separately, so is closer in some regards to a flow chart.

Design Recipes for FPGAs. http://dx.doi.org/10.1016/B978-0-08-097129-2.00022-2

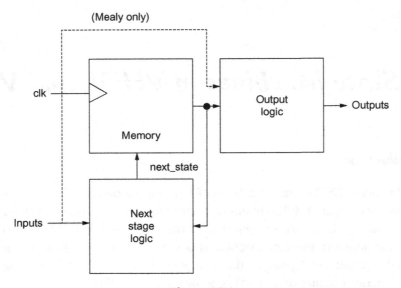

Figure 22.1
Hardware state machine structure.

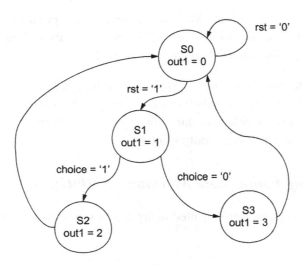

Figure 22.2
State transition diagram.

```vhdl
      clk, rst, choice : in std_logic;
      count : out std_logic
   );
end entity fsm;
architecture simple of fsm1 is
      type state_type is ( s0, s1, s2, s3 );
      signal current, next_state : state_type;
begin
  process ( clk )
  begin
    if ( clk = 1 ) then
      current <= next_state;
    end if;
  end process;

  process ( current )
  begin
    case current is
      when s0 =>
        out <= 0;
        if ( rst = 1) then
          next <= s1;
        else
          next <= s0;
        end if;
      when s1=>
        out <= 1;
        if ( choice = 1) then
          next <= s3;
        else
          next <= s2;
        end if;
      when s2=>
        out <= 2;
        next <= s0;
      when s3=>
        out <= 3;
        next <= s0;
    end case;
  end process;
end;
```

It must be noted that not all state machines will neatly have a number of states exactly falling to a power of 2, and so unused states must also be managed using the "when-others" approach described elsewhere in this book.

It is also the case that the two processes can be combined into a single process, which can reduce the risk of glitches being introduced by the synthesis tools, especially from incorrect assignments. It can also make debugging simpler.

22.4 *Implementing Finite State Machines in Verilog*

The transition diagram can also be implemented in Verilog, again using two case statements (one for the state transitions and one for the outputs):

```verilog
module fsm (
  count, // Output Value
  clk,  // Clock
  rst,  // Reset
  choice // Decision /Choice Value
  );


output [1:0] count;
input clk;
input rst;
input choice;

reg [1:0] count;

reg [1:0] state; // state variable

parameter s0=0, s1=1, s2=2, s3=3;

always @(state)
  begin
    case (state)
    s0:
      count = 2'b00;
    s1:
      count = 2'b01;
    s2:
      count = 2'b10;
    s3:
      count = 2'b11;
    default:
      count = 2'b00;
  endcase
end

always @(posedge clk or posedge rst)
begin
  if (rst == 0)
    state = s0;
  else
    case (state)
      s0:
        state = s1;
      s1:
        if (choice)
          state = s3;
        else
          state = s2;
```

```
      s2:
        state = s0;
      s3:
        state = s0;
    endcase
end

endmodule
```

22.5 Testing the Finite State Machine Model

Whether the VHDL or Verilog is being used, a test bench is required to evaluate the behavior and check that it is correct. In this example, Verilog has been used, but the principle is the same for both HDLs. The idea is to reset the FSM, then clock through the sequence with choice = 0 (which will go from state S0 to S1 and then S2, returning to S0), and then the choice is set to 1, and this time the sequence will go from state S0 to S1 and then S3, returning to S0. The output variable (*counter_output*) shows the state number as an output. The simulation is shown in Figure 22.3.

22.6 Summary

Finite State Machines are a fundamental technique for designing control algorithms in digital hardware. This chapter of this book is purely an introduction to the key concepts and if the reader is not already fully familiar with the basic concepts of digital hardware design, you are encouraged to obtain a standard text on digital design techniques to complement the practical implementation methods described in this book.

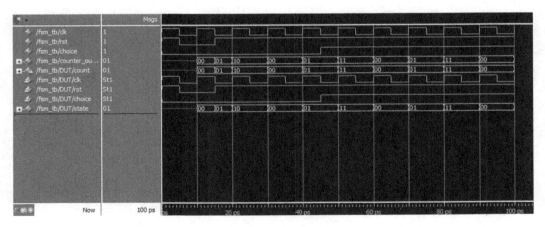

Figure 22.3
FSM simulation results.

CHAPTER 23

Fixed Point Arithmetic

23.1 Introduction

In digital systems and HDLs such as VHDL or Verilog we have access to a range of types from bits and Booleans (which consist of two states 1 and 0, (or true and false, respectively) which are effectively enumerated types, through integer numbers (including positive and natural subtypes), and eventually we can use real numbers (floating point). Unfortunately, the big drawback is not necessarily what we can use in a particular HDL, but rather what we can synthesize in hardware. In most cases it is still not possible to directly synthesize real numbers directly.

Despite recent research efforts and standardization efforts, there is still a limited availability of packages and libraries that support both fixed point and floating point arithmetic specifically for FPGAs. If we consider most FPGA applications, there is a need for some DSP type applications, and generally a form of fixed point arithmetic will be adequate in most of these cases. While there are now some additional packages available as a result of the inclusion of the numeric_std library into VHDL, it is nevertheless useful to understand how the number systems work when translated into specific forms for use in digital systems.

So, what is fixed point arithmetic and how can we use it in FPGA design? In integer arithmetic, whether unsigned, signed, or std_logic, the basis of the number is a bitwise representation of an integer number, with no decimal point. For example, to represent the number 23, using 8 bits, we simply set a bit for each binary element required to construct the integer value of 23. This is shown in Figure 23.1.

If we require a negative number, then we use the *signed* approach, where the MSB is simply the sign bit as shown in Figure 23.2. In fact, the two's complement notation (discussed in Chapter 22), can be obtained by inverting the bits and adding one to the LSB. We have in fact discussed the difference between the unsigned and signed format elsewhere in this book; however, in summary the signed numbers use a bit to denote the sign of the number (positive or negative), at the expense of 1 bit of range (effectively halving the maximum magnitude, but allowing positive or negative values). Unsigned numbers in contrast only represent the magnitude.

Design Recipes for FPGAs. http://dx.doi.org/10.1016/B978-0-08-097129-2.00023-4

With this basic idea of handling numbers, we can extend the notation to a *fixed point* scheme by defining where the decimal point will go. For example, in the same number scheme shown we have 8 bits. We can therefore define this in terms of 5 bits above the decimal point and 3 below it. This will give some limited fractional usage for the numbers. The way that this is implemented is by using fractions of 1 for each negative bit to the right of the decimal point. As an example, take the same number in terms of bits used in Figure 23.3 and use the new fixed point numbering system for the bits. In this case, we get a value of −2.875.

The nice thing about this notation is that the bitwise functions defined for the integer-based ALU developed previously can also be applied to this new fixed point notation with almost no modification. The only difference is that we need to translate from the new fixed point type to a std_logic_vector type in VHDL and also consider how to handle overflow conditions.

128	64	32	16	8	4	2	1
0	0	0	1	0	1	1	1

16 + 4 + 2 + 1 = 23

Figure 23.1
Basic binary notation.

−128	64	32	16	8	4	2	1
1	1	1	0	1	0	0	1

−128 + 64 + 32 + 8 + 1 = −23

Figure 23.2
Negative number binary notation.

−16	8	4	2	1	½	¼	⅛
1	1	1	0	1	0	0	1

−16 + 8 + 4 + 1 + 0.125 = −2.875

Figure 23.3
Fixed point notation.

For example, if two numbers are added together and the result is too large, how is this handled by the fixed point algorithms? Do we simply flag an overflow and output the result? Or do we set the maximum value and output this?

Similarly, for numbers which may be too small and for which we can potentially lose precision, do we simply round up or flag another loss of precision condition? These are questions that the designer needs to answer for their application, but for the rest of the chapter a simple approach will be taken that illustrates how the basic functions operate, and the details of handling these issues will be left to the reader, unless specifically identified and discussed.

In practice, it is useful for the designer to establish basic rules for whichever number scheme is used, such as aligning decimal points for certain operations, such as addition, or using automatic vector size increase for multiplication to ensure there are always an adequate number of bits available to handle the maximum possible range of numbers.

In the next section, the construction of a basic numeric package is discussed; however, it should be reiterated that the user can take advantage of the numeric_std package in VHDL.

23.2 Basic Fixed Point Types in VHDL

The first task in defining a custom fixed point library is to specify a new type for the numbers. The closest similar types in standard VHDL that can be synthesized are unsigned and signed. These are defined in terms of a specific number of bits. In most cases we are interested in linking directly to std_logic systems, and so in this case we can effectively define a new type based on an array of std_logic bits. For the remainder of this chapter we will discuss signed arithmetic only, as this is the most potentially useful from a DSP and application point of view.

The basic VHDL type that defines our base type is to be called *fixsign* and is defined as an unrestricted array of std_logic:

```
1    type fixsign is array ( integer range <> ) of std_logic;
```

From this, we can define specific subtypes that have a defined range of fixed point. For example, we can define a type (this is often called a *fractional integer* type) that has 8 bits above the decimal point and 3 bits below (for example, 00000001.001) using the following declaration:

```
1    Subtype fp8_3 is fixsign ( 8 downto -3);
```

With this, we can declare signals of this new type and use them in fixed point VHDL models:

```
1    signal a1 : fp8_3;
2    a1 <= XOCA;
```

Clearly this is useful but limited, as this type needs to be able to be converted from one type to another easily and quickly. The simplest way to manage this process is to create a new package that contains not only the type declarations, but also the functions that are associated with this set of types. Therefore we can define a new package called fp_pkg that, as a minimum, contains these type declarations:

```
1    package fp_pkg is
2            type fixsign is array (integer range <>) of std_logic;
3            subtype fp8_3 is fixsign ( 8 downto −3);
4    end package;
5
6    package body fp_pkg is
7    -- The contents of the package go here
8    end package body;
```

We can now use this package in a VHDL model by compiling the package into the current work library and calling the package as we need it. We can also create a new library so that the package could be used more generally.

```
1    Use work.fp_pkg.all;
```

This will provide access to all the fixed point functions and types required. In this library, we have two types of functions. The first type is required for translating physical types (such as std_logic_vector) to our new types and vice versa. These are important as they will be synthesized and eventually end up on hardware. The second type are purely for debug purposes and displaying values. For example, it is useful to be able to convert fixed point data to real numbers and then use the real'image VHDL function to display the value. This could be extremely useful for debugging where a behavioral model *would* be able to represent real numbers, and so this could form a very helpful way to establish both the accuracy and validity of the "digital" equivalent system. A useful set of functions to facilitate this is therefore presented in this chapter. Again, these are exemplar functions, and readers are encouraged to develop these basic functions and produce their own for their own applications.

23.3 Fixed Point Functions in VHDL

23.3.1 Fixed Point to STD_LOGIC_VECTOR Functions

The most important functions are the conversion between fixed point and std_logic_vector variables. If we can translate from one to the other, then we can use our standard logic functional blocks, where appropriate, on the fixed point data directly, rather than needing to come up with brand-new blocks every time. The easiest function is the mapping from fixed point to std_logic_vector and is simply a matter of starting from the LSB defined in the range of the fixed point number and then setting each bit on the output std_logic_vector in turn to the correct value. The VHDL for this is given as follows:

```
1    function fp2std_logic_vector (d:fixsign;top:integer;low:integer)
2    return std_logic_vector is
3      variable outval : std_logic_vector ( top−low downto 0 ) := (others => '0');
```

```
4    begin
5      for i in 0 to top-low loop
6        outval(i) := d(i+low);
7      end loop;
8    return outval;
9    end;
```

If we look at this function we can see that the arguments to the function are the fixed point number, and then the two integer values that denote the number of bits above and below the decimal point, respectively. For example, if our notation is 8.3, the function call in this case would be:

```
1    q <= fp2std_logic_vector(d,8,-3);
```

Notice the negative number denoting the bits below the decimal point. If you would prefer both numbers to be positive, they can simply be changed. One reason for using the negative form is that the numbers match the basic type definition and therefore make checking easy.

Similarly, we can convert from std_logic_vector back to fixed point using a very similar function in the opposite direction:

```
1    function std_logic_vector2fp
2    (d:std_logic_vector;top:integer;low:integer)
3    return fixsign is
4      variable outval : fixsign ( top downto low ) := (others => '0');
5    begin
6      for i in 0 to top-low loop
7        outval(i+low) := d(i);
8      end loop;
9    return outval;
10   end;
```

with the similar usage:

```
1    q <= std_logic_vector(d,8,-3);
```

Using these functions, the conversion between the std_logic_vector and fixed point arithmetic domains becomes straightforward. Also, these functions are easily synthesizable as they simply map bits and do not carry out any sophisticated functions other than that.

23.3.2 Fixed Point to Real Conversion

An extremely useful function is the ability to convert from fixed point to real numbers. Obviously this has no use for synthesis, but is ideal for adding checking and reports to test benches. As a result we only define a single function *fp2real* which takes a fixed point number and converts to a real number for display. Once we have the number, then the real'image function can be used to display the value. The VHDL for the conversion function is given here:

```
1    function fp2real (d:fixsign; top:integer; low:integer)
2    return real is
```

```
3      variable outreal : real := 0.0;
4      variable mult : real := 1.0;
5      variable max : real := 1.0;
6      variable debug : boolean := false;
7    begin
8    for i in 0 to top-1 loop
9      if d(i) = '1' then
10       outreal := outreal + mult;
11       if debug then
12         report " fp2real : " & integer'image(i);
13       end if;
14     end if;
15     mult := mult * 2.0;
16   end loop;
17   if debug then
18     REPORT " fp2real middle : " & real'image(outreal);
19   end if;
20   max := mult;
21   mult := 0.5;
22   for i in -1 downto low loop
23     if d(i) = '1' then
24       outreal := outreal + mult;
25       if debug then
26         report " fp2real : " & integer'image(i);
27       end if;
28     end if;
29     mult := mult * 0.5;
30   end loop;
31   if debug then
32     REPORT " fp2real : " & real'image(outreal);
33   end if;
34
35   if d(top) = '1' then
36     outreal := outreal - max;
37   end if;
38   if debug then
39     REPORT " fp2real FINAL VALUE : " & real'image(outreal);
40   end if;
41
42   return outreal;
43   end;
```

This function is a simple converter that handles the bits above and below the decimal point in turn. Also notice the internal Boolean debug variable that allows checking of each individual bit. This can be very useful when observing the passing of numbers across boundaries ensuring correct translation; this defaults to false (off). If we need to report a fixed point value, we can therefore use this function to report the values using simple VHDL such as this:

```
1    d : fp8_3;
2    dr : real;
3    dr <= fp2real(fp8_3,8,-3);
4    report "The value is : " & real'image(dr);
```

23.4 Testing the VHDL Fixed Point Functions

As stated previously, we can use these functions to incorporate standard std_logic ALU functions into the model. In this simple test case, we are using the standard n-bit adder created in Chapter 21 on ALUs to add two fixed point numbers together. How does this work? What we do is convert the two input fixed point numbers into std_logic_vectors, apply them to the adder block, then convert the output back to a fixed point number. We can convert both inputs and output into real numbers for observation on the screen. Note that although the representation of numbers is different, the values are the same. Also, this is a low-level approach that could be carried out using synthesis of the numeric_std equivalent types.

```vhdl
1    library ieee;
2    use ieee.std_logic_1164.all;
3    use ieee.numeric_std.all;
4
5    use work.fp_pkg.all;
6
7    entity simple1 is
8    end entity simple1;
9
10   architecture tb of simple1 is
11     signal clk : std_logic := '0';
12     signal cin : std_logic := '0';
13     signal cout : std_logic;
14     signal testa : fp8_3 := "000000000000";
15     signal testa1 : fixsign ( 8 downto -3 );
16     signal testa2 : fixsign ( 8 downto -3 );
17     signal testb1 : fixsign ( 8 downto -3 );
18     signal testsum : fixsign ( 8 downto -3 );
19     signal as : signed ( 11 downto 0) := X"000";
20     signal a1std : std_logic_vector ( 11 downto 0) := X"800";
21     signal b1std : std_logic_vector ( 11 downto 0) := X"800";
22     signal sum : std_logic_vector ( 11 downto 0) ;
23     signal a1out : real;
24     signal b1out : real;
25     signal a2out : real;
26     signal sumout : real;
27     signal a1 : integer := 0;
28     signal bs : signed ( 11 downto 0) := X"8f0";
29
30     component add_beh
31     generic (
32       top : integer := 7
33     );
34     port (
35       signal a : in std_logic_vector(top downto 0);
36       signal b : in std_logic_vector(top downto 0);
37       signal cin : in std_logic;
38       signal cout : out std_logic;
39       signal sum : out std_logic_vector(top downto 0)
40     );
41     end component;
```

```
42        for all : add_beh use entity work.add_beh;
43
44
45    begin
46      clk <= not clk after 1 us;
47
48      DUT :add_beh generic map ( 11 ) port map ( a1std, b1std, cin, cout, sum);
49
50      p1 : process (clk)
51      begin
52        as <= as + 1;
53        testa1 <= signed2fp(as,8,-3);
54        testb1 <= signed2fp(bs,8,-3);
55        a1out <= fp2real(testa1,8,-3);
56        b1out <= fp2real(testb1,8,-3);
57        a1std <= fp2std_logic_vector(testa1,8,-3);
58
59        b1std <= fp2std_logic_vector(testb1,8,-3);
60        testa2 <= std_logic_vector2fp(a1std,8,-3);
61        testsum <= std_logic_vector2fp(sum,8,-3);
62        a2out <= fp2real(testa2,8,-3);
63        sumout <= fp2real(testsum,8,-3);
64        report "a1out : " & real'image(a1out);
65        report "a2out : " & real'image(b1out);
66        report "sumout : " & real'image(sumout);
67      end process p1;
68    end;
```

An important aspect to note in this model is the use of signals and a clock (clk). By making this model synchronous, we have ensured correct, predictable behavior, but on each clock cycle there are several delays built in. The final observed result on sumout (the real number output for display) will appear 2 clock cycles after the data is input to the model.

In this case we are using signed numbers as the original input (as) as these can be incremented easily and setting one number to a constant (bs). These inputs are converted to real numbers (a1out, b1out) that are displayed to the screen to show the results.

23.5 Fixed Point Types in Verilog

Verilog has built-in types for signed and unsigned when we define registers (reg). The default type for a reg is *unsigned*, and so, therefore, we can define a 16-bit unsigned integer using the following syntax:

```
1    reg [15:0] unsigned_integer;
```

It can be a little unclear as to where the decimal point occurs, and one approach to make this explicit in the declaration is to offset the array indices accordingly. For example, to use an unsigned integer of the form 8:8, the declaration could be redefined as follows:

```
1    reg [7:-8] unsigned_integer;
```

Signed integers are defined in Verilog in exactly the same manner, except using the addition of the keyword *signed*, so taking the previous example, the declaration would become:

```
1    reg signed [7:-8] signed_integer;
```

When using these numbers, no additional functions are required; however, care needs to be taken when shifting to ensure that the correct notation is used (i.e., that it is consistent), and the result is shifted by the appropriate number of bits to maintain the correct accuracy.

One of the nice aspects of Verilog is that it is not really necessary to define new types and conversion routines as with VHDL, and it is possible to allow Verilog to handle the conversions. For example, the low level definition of integers or fixed point data types as registers means that the conversion between the two is implicit, as long as the number of bits is consistent, making conversions very simple.

23.6 Floating Point Types in Verilog

As for VHDL, there are standard type definitions based on IEEE-754 or IEEE-854 for generic floating point (real) types; however, while Verilog supports the use of real types, synthesis is not possible directly. The user has two options in order to implement these types in Verilog: to define specific composite types (similar to fixed point) and associated functions. However, most of the FPGA vendors also offer "blocks" that can provide DSP functions based on floating point types, effectively as IP blocks.

The floating point types can be defined as single or double precision, where the single precision consists of a single sign bit, 8 bits for the exponent and the remaining 23 bits for the mantissa. Double precision also has a single sign bit, 11 bits for the exponent and 52 for the mantissa.

In addition to the basic number format, IEEE Std 754 also defines a number of rounding methods, and it is important to ensure that not only should the number types be defined, but also the rounding and arithmetic functions.

23.7 Summary

This chapter has introduced the concept of fixed point arithmetic in VHDL and provided a basic package of functions and types to get started using VHDL. It must be stressed that this package is purely for exemplar designs and the reader is encouraged to either use commercially available libraries for optimum performance or to develop their own libraries. The usage of fixed point functions in Verilog is also briefly introduced.

Floating point functionality has also been introduced; however, the overhead in the FPGA is such that unless double precision is applied, the accuracy will be limited due to rounding errors, and therefore fixed point is still usually quite acceptable for most practical applications. In practice, as the size of FPGAs has increased to the extent that a microprocessor core can be quite easily implemented on the FPGA, in most cases it is simply a case of writing the floating point code in C and running a processor core directly on the FPGA with the high-level code on the core.

CHAPTER 24

Counters

24.1 Introduction

One of the most commonly used applications of flip-flops in practical systems is counters. They are used in microprocessors to count the program instructions (program counter or PC), for accessing sequential addresses in memory (such as ROM) or for checking progress of a test. Counters can start at any value, although most often they start at zero and they can increment or decrement. Counters may also change values by more than one at a time, or in different sequences, such as grey code, binary, or binary coded decimal (BCD) counters.

24.2 Basic Binary Counter using VHDL

The simplest counter to implement in many cases is the basic binary counter. The basic structure of a counter is a series of flip-flops (a register), that is controlled by a reset (to reset the counter to zero literally) and a clock signal used to increment the counter. The final signal is the counter output, the size of which is determined by the generic parameter n, which defines the size of the counter. The symbol for the counter is given in Figure 24.1. Notice that the reset is active low and the counter and reset inputs are given in a separate block of the symbol as defined in the IEEE standard format.

From an FPGA implementation point of view, the value of generic n also defines the number of D type flip-flops required (usually a single LUT) and hence the usage of the resources in the FPGA. A simple implementation of such a counter is given here:

```
1    library ieee;
2    use ieee.std_logic_1164.all;
3    use ieee.numeric_std.all;
4
5    entity counter is
6      generic (
7              n : integer := 4
8          );
9      port (
10            clk : in std_logic;
11            rst : in std_logic;
12          output : out std_logic_vector((n-1) downto 0)
13      );
14    end;
```

Design Recipes for FPGAs. http://dx.doi.org/10.1016/B978-0-08-097129-2.00024-6

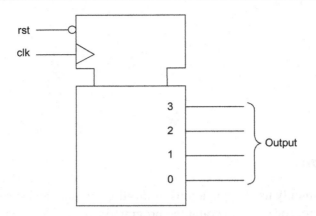

Figure 24.1
Simple binary counter.

```
15
16    architecture simple of counter is
17    begin
18      process(clk, rst)
19        variable count : unsigned((n−1) downto 0);
20      begin
21        if rst = '0' then
22          count := (others => '0');
23        elsif rising_edge(clk) then
24          count := count + 1;
25        end if;
26        output <= std_logic_vector(count);
27      end process;
28    end;
```

The important aspect of this approach to the counter VHDL is that this is effectively a state machine; however, we do not have to explicitly define the next state logic, as this will be taken care of by the synthesis software. This counter can now be tested using a simple test bench that resets the counter and then clocks the state machine until the counter flips round to the next counter round. The test bench is given as follows:

```
1     library ieee;
2     use ieee.std_logic_1164.all;
3     use ieee.numeric_std.all;
4
5     entity CounterTest is
6     end CounterTest;
7
8     architecture stimulus of CounterTest is
9           signal rst : std_logic := '0';
10          signal clk : std_logic:='0';
11          signal count : std_logic_vector (3 downto 0);
12
13          component counter
```

```
14                  port(
15                      clk : in std_logic;
16                      rst : in std_logic;
17                      output : out std_logic_vector(3 downto 0)
18                  );
19              end component;
20              for all : counter use entity work.counter ;
21
22      begin
23          DUT: counter port map(clk=>clk,rst=>rst,output=>count);
24          clk <= not clk after 1 us;
25          process
26          begin
27            rst<='0','1' after 2.5 us;
28            wait;
29          end process;
30      end;
```

Using this simple VHDL test bench, we reset the counter until 2.5 μs and then the counter will count on the rising edge of the clock after 2 μs (i.e., the counter is running at 500 kHz).

If we dissect this model, there are several interesting features to notice. The first is that we need to define an internal variable count rather than simply increment the output variable q. The output variable q has been defined as a standard logic vector (std_logic_vector) and with it being defined as an output we cannot use it as the input variable to an equation. Therefore we need to define a local variable (in this case, count) to store the current value of the counter.

The initial decision to make is whether to use a variable or a signal. In this case, we need an internal variable that we can effectively treat as a sequential signal, and also one that changes instantaneously, which immediately requires the use of a variable. If we chose a signal, then the changes would only take place when the cycle is resolved (i.e., the next time the process is activated).

The second decision is what type of unit to use for the count variable. The output variable is a std_logic_vector type, which has the advantage of being an array of std_logic signals, and so we don't need to specify the individual bits on a word; this is done automatically. The major disadvantage, however, is that the std_logic_vector does not support simple arithmetic operations, such as addition, directly. In this example, we want the counter to have a simple definition in VHDL and so the best compromise type that has the bitwise definition and also the arithmetic functionality would be the unsigned or signed type. In this case, we wish to have a direct mapping between the std_logic_vector bits and the counter bits, so the unsigned type is appropriate. Thus the declaration of the internal counter variable count is as follows:

```
1       variable count : unsigned((n−1) downto 0);
```

The final stage of the model is to assign the internal value of the count variable to the external std_logic_vector q. Luckily, the translation from unsigned to std_logic_vector is fairly direct, using the standard casting technique:

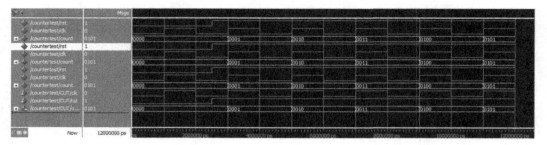

Figure 24.2
Simple binary counter simulation in VHDL.

```
1    q <= std_logic_vector(count);
```

As the basic types of both q and count are consistent, this can be done directly. The resulting model simply counts up the binary values on the clock edge as specified, as is shown in Figure 24.2.

24.3 Simple Binary Counter using Verilog

With the same basic specification as the VHDL counter, it is possible to implement a basic counter in Verilog using the same architecture of the model.

```
1    module counter (
2      clk,                    // clock input
3      rst,                    // reset (active low)
4      counter_output   // counter output
5      );
6
7    input clk;
8    input rst;
9
10   output [3:0] counter_output;
11
12   wire clk;
13   wire rst;
14
15   reg [3:0] counter_output ;
16
17   always @ (posedge clk)
18   begin : count
19     if (rst == 1'b0) begin
20       counter_output <= #1 4'b0000;
21     end
22     else begin
23       counter_output <= #1 counter_output + 1;
24     end
25   end
```

```
26
27      endmodule
```

The model has the same connection points and operation, and will be treated in the same way for synthesis by the design software. The test bench is slightly different from the VHDL one in that it is much more explicit about the test function as well as the functionality of the test bench. For example, if we look at the top of the test bench Verilog has the $display and $monitor commands, which enable the time and variable values to be displayed in the monitor of the simulation as well as looking at the waveforms.

```
1      $display ("time\t clk reset counter");
2      $monitor ("%g\t %b %b %b",
3      $time, clk, rst, counter_output);
```

Using this test bench and Verilog model the behavior of the simulation can also be verified as shown in Figure 24.3.

24.4 Synthesized Simple Binary Counter

At this point it is useful to consider what happens when we synthesize this VHDL, so to test this point the VHDL model of the simple binary counter was run through a typical RTL synthesis software package (Leonardo Spectrum) with the resultant synthesized VHDL model given here:

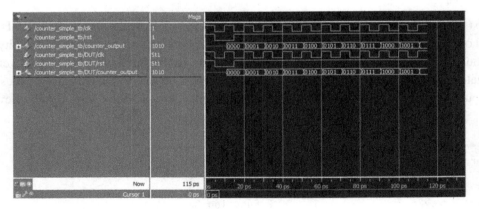

Figure 24.3
Simple binary counter simulation in Verilog.

```
1    entity counter is
2      port (
3        clk : IN std_logic ;
4        rst : IN std_logic ;
5        output : OUT std_logic_vector (3 DOWNTO 0)) ;
6    end counter ;
7
8    architecture simple of counter is
9      signal clk_int, rst_int, output_dup0_3, output_dup0_2, output_dup0_1,
10       output_dup0_0, output_nx4, output_nx7, output_nx10, NOT_rst,
11       output_NOT_a_0: std_logic ;
12
13   begin
14     output_obuf_0 : OBUF port map ( O=>output(0), I=>output_dup0_0);
15     output_obuf_1 : OBUF port map ( O=>output(1), I=>output_dup0_1);
16     output_obuf_2 : OBUF port map ( O=>output(2), I=>output_dup0_2);
17     output_obuf_3 : OBUF port map ( O=>output(3), I=>output_dup0_3);
18     rst_ibuf : IBUF port map ( O=>rst_int, I=>rst);
19     output_3_EXMPLR_EXMPLR : FDC port map ( Q=>output_dup0_3, D=>output_nx4,
20       C=>clk_int, CLR=>NOT_rst);
21     output_2_EXMPLR_EXMPLR : FDC port map ( Q=>output_dup0_2, D=>output_nx7,
22       C=>clk_int, CLR=>NOT_rst);
23     output_1_EXMPLR_EXMPLR : FDC port map ( Q=>output_dup0_1, D=>output_nx10,
24       C=>clk_int, CLR=>NOT_rst);
25     output_0_EXMPLR_EXMPLR : FDC port map ( Q=>output_dup0_0, D=>
26       output_NOT_a_0, C=>clk_int, CLR=>NOT_rst);
27     clk_ibuf : BUFGP port map ( O=>clk_int, I=>clk);
28     output_nx4 <= (not output_dup0_3 and output_dup0_2 and output_dup0_1 and
29      output_dup0_0) or (output_dup0_3 and not output_dup0_0) or (output_dup0_3
30       and not output_dup0_2) or (output_dup0_3 and not output_dup0_1) ;
31     output_nx7 <= (output_dup0_2 and not output_dup0_0) or (not output_dup0_2
32       and output_dup0_1 and output_dup0_0) or (output_dup0_2 and not
33      output_dup0_1) ;
34     output_nx10 <= (output_dup0_0 and not output_dup0_1) or (not
35      output_dup0_0 and output_dup0_1) ;
36     NOT_rst <= (not rst_int) ;
37     output_NOT_a_0 <= (not output_dup0_0) ;
38   end simple ;
```

The first obvious aspect of the model is that it is much longer than the simple RTL VHDL created originally. The next stage logic is now in evidence; as this is synthesized, the physical gates must be defined for the model. Finally the outputs are buffered, which leads to even more gates in the final model. If the optimization report is observed, the overall statistics of the resource usage of the FPGA can be examined (in this case, a Xilinx Virtex-II Pro device):

Cell	Library	References		Total Area
BUFGP	xcv2p	1 x	1	1 BUFGP
FDC	xcv2p	4 x	1	4 Dffs or Latches
IBUF	xcv2p	1 x	1	1 IBUF
LUT1	xcv2p	2 x	1	2 Function Generators
LUT2	xcv2p	1 x	1	1 Function Generators
LUT3	xcv2p	1 x	1	1 Function Generators

```
LUT4     xcv2p     1 x     1      1 Function Generators
OBUF     xcv2p     4 x     1      4 OBUF

 Number of ports :                      6
 Number of nets :                      17
 Number of instances :                 15
 Number of references to this view :    0
Total accumulated area :
 Number of BUFGP :                      1
 Number of Dffs or Latches :            4
 Number of Function Generators :        5
 Number of IBUF :                       1
 Number of OBUF :                       4
 Number of gates :                      5
 Number of accumulated instances :     15
Number of global buffers used: 1
************************************************
Device Utilization for 2VP2fg256
************************************************
Resource              Used    Avail   Utilization
-------------------------------------------------
IOs                     5     140      3.57%
Global Buffers          1      16      6.25%
Function Generators     5    2816      0.18%
CLB Slices              3    1408      0.21%
Dffs or Latches         4    3236      0.12%
Block RAMs              0      12      0.00%
Block Multipliers       0      12      0.00%
```

In this simple example, it can be seen that the overall utilization of the FPGA is minimal, with the relative resource allocation of IOs, buffers and functional blocks. This is an important aspect of FPGA design in that, even though the overall device may be underutilized, a particular resource (such as IO) might be used up. The output VHDL can then be used in a physical place and route software tool (such as the Xilinx Design Navigator) to produce the final bit file that will be downloaded to the device.

24.5 Shift Register

While a shift register is, strictly speaking, not a counter, it is useful to consider this in the context of other counters as it can be converted into a counter with very small changes. We will consider this element layer in this book, in more detail, but consider a simple case of a shift register that takes a single bit and stores in the least significant bit of a register and shifts each bit up one bit on the occurrence of a clock edge. If we consider an n-bit register and show the status before and after a clock edge, then the functionality of the shift register becomes clear, as shown in Figure 24.4.

A basic shift register can be implemented in VHDL as shown here:

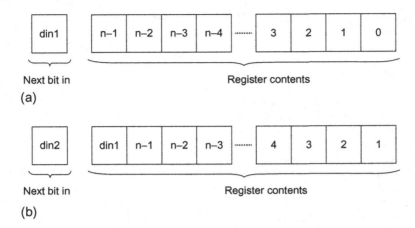

Figure 24.4
Simple shift register functionality: (a) Before the clock edge; (b) After the clock edge.

```
1    library ieee;
2    use ieee.std_logic_1164.all;
3
4    entity shift_register is
5      generic (
6          n : integer := 4;
7      port (
8          clk : in std_logic;
9          rst : in std_logic;
10         din : in std_logic;
11         q : out std_logic_vector((n-1) downto 0)
12     );
13    end entity;
14
15    architecture simple of shift_register is
16    begin
17      process(clk, rst)
18        variable shift_reg : std_logic_vector((n-1) downto 0);
19      begin
20        if rst = '0' then
21          shift_reg := (others => '0');
22        elsif rising_edge(clk) then
23          shift_reg := shift_reg(n-2 downto 0) & din;
24        end if;
25        q <= shift_reg;
26      end process;
27    end architecture simple;
```

The interesting parts of this model are very similar to the simple binary counter, but subtly different. As for the counter, we have defined an internal variable (shift_reg), but unlike the counter we do not need to carry out arithmetic functions, so we do not need to define this as an unsigned variable. Instead, we can define directly as a std_logic_vector, the same as the output q.

Notice that we have an asynchronous clock in this case. As we have discussed previously in this book, there are techniques for completely synchronous sets or resets, and these can be used if required.

The fundamental difference between the counter and the shift register is in how we move the bits around. In the counter we use arithmetic to add one to the internal counter variable (count). In this case, we just require shifting the register up by one bit, and to achieve this we simply assign the lowest $(n - 1)$ bits of the internal register variable (shift_reg) to the upper $(n - 1)$ bits and concatenate the input bit (din), effectively setting the lowest bit of the register to the input signal (din). This can be accomplished using the VHDL following:

```
1    shift_reg := shift_reg(n-2 downto 0) & din;
```

The final stage of the model is similar to the basic counter in that we then assign the output signal to the value of the internal variable (shift_reg) using a standard signal assignment. In the shift register, we do not need to cast the type as both the internal and signal variable types are std_logic_vector:

```
1    q <= shift_reg;
```

We can also implement the shift register in Verilog, with the listing as shown here:

```
1    module shift_register (
2      clk,                   // clock input
3      rst,                   // reset (active low)
4      din,                   // Digital Input
5      shiftreg         // shift register
6      );
7
8    input clk;
9    input rst;
10   input din;
11
12   output [7:0] shiftreg;
13
14   wire clk;
15   wire rst;
16   wire din;
17
18   reg [7:0] shiftreg ;
19
20   always @ (posedge clk)
21   begin : count
22     if (rst == 1'b0) begin
23       shiftreg <= #1 4'b00000000;
24     end
25     else begin
26       shiftreg <= #1 {din, shiftreg[7:1]};
27     end
28   end
29
30   endmodule
```

In both cases (VHDL and Verilog) we can test the behavior of the shift register by applying a data sequence and observing the shift register variable in the model, and in the case of the Verilog we can also add a $monitor command to display the transitions as they happen in the transcript of the simulator. The Verilog test bench code is given as:

```verilog
module shift_register_tb();
// declare the signals
reg clk;
reg rst;
reg din;
wire [7:0] shift_register_values;

// Set up the initial variables and reset
initial begin
  $display ("time\t clk reset counter");
  $monitor ("%g\t %b %b %b %h",
  $time, clk, rst, din, shift_register_values);
  clk = 1;    // initialize the clock to 1
  rst = 1;    // set the reset to 1 (not reset)
  din = 0;           // Initialize the digital input
  #5 rst = 0; // reset = 0 : resets the counter
  #10 rst = 1; // reset back to 1 : counter can start
  #4 din = 0; // test data sequence starting at cycle time 16
  #10 din = 1; // din = 1 test data sequence
  #10 din = 0; // din = 0 test data sequence
  #10 din = 0; // din = 0 test data sequence
  #10 din = 1; // din = 1 test data sequence
  #10 din = 1; // din = 1 test data sequence
  #10 din = 0; // din = 0 test data sequence
  #10 din = 1; // din = 1 test data sequence
  #10 din = 0; // din = 0 test data sequence
  #10 din = 1; // din = 1 test data sequence
  #1000 $finish; // Finish the simulation
end

// Clock generator
always begin
  #5 clk = ~clk; // Clock every 5 time slots
end

// Connect DUT to test bench
shift_register DUT (
clk,
rst,
din,
shift_register_values
);

endmodule
```

The resulting simulation of the shift register can be seen in Figure 24.5.

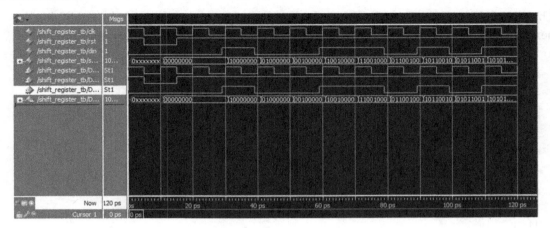

Figure 24.5
Simple shift register simulation.

24.6 The Johnson Counter

The Johnson counter is a counter that is a simple extension of the shift register. The only difference between the two is that the Johnson counter has its least significant bit inverted and fed back into the most significant bit of the register. In contrast to the classical binary counter with 2n states, the Johnson counter has 2^n states. While this has some specific advantages, a disadvantage is that the Johnson counter has what is called a *parasitic counter* in the design. In other words, while the 2^n counter is operating, there is another state machine that also operates concurrently with the Johnson counter using the unused states of the binary counter. A potential problem with this counter is that if, due to an error, noise or other glitch, the counter enters a state NOT in the standard Johnson counting sequence, it cannot return to the correct Johnson counter without a reset function. The normal Johnson counter sequence is shown in the following table:

Count	Q(3:0)
0	0000
1	1000
2	1100
3	1110
4	1111
5	0111
6	0011
7	0001

The VHDL implementation of a simple Johnson counter can then be made by modifying the next stage logic of the internal shift_register function as shown in the following listing:

```
1     library ieee;
2     use ieee.std_logic_1164.all;
3
4     entity johnson_counter is
5       generic (
6            n : integer := 4;
7       port (
8            clk : in std_logic;
9            rst : in std_logic;
10            din : in std_logic;
11           q : out std_logic_vector((n-1) downto 0)
12       );
13    end entity;
14
15    architecture simple of Johnson_counter is
16    begin
17      process(clk, rst)
18        variable j_state : std_logic_vector((n-1) downto 0);
19      begin
20        if rst = '0' then
21          j_state:= (others => '0');
22        elsif rising_edge(clk) then
23          j_state:= not j_state(0) & j_state(n-1 downto 1);
24        end if;
25        q <= j_state;
26      end process;
27    end architecture simple;
```

Notice that the concatenation is now putting the inverse (NOT) of the least significant bit of the internal state variable (j_state(0)) onto the next state most significant bit, and then shifting the current state down by one bit.

It is also worth noting that the counter does not have any checking for the case of an incorrect state. It would be sensible in a practical design to perhaps include a check for an invalid state and then reset the counter in the event of that occurrence. The worst-case scenario would be that the counter would be incorrect for a further 7 clock cycles before correctly resuming the Johnson counter sequence.

In a similar manner we can implement a Johnson counter in Verilog using the code given here:

```
1     module johnson_counter (
2       clk,                  // clock input
3       rst,                  // reset (active low)
4       johnsonreg            // shift register
5       );
6
7     input clk;
8     input rst;
9
10    output [7:0] johnsonreg;
```

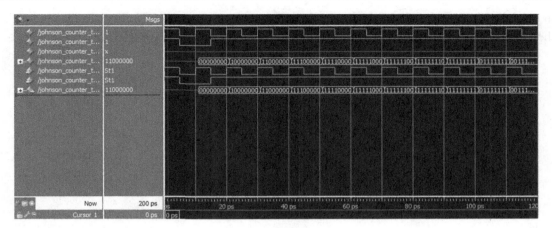

Figure 24.6
Johnson counter simulation.

```
11
12    wire clk;
13    wire rst;
14
15    reg [7:0] johnsonreg ;
16
17    always @ (posedge clk)
18    begin : count
19      if (rst == 1'b0) begin
20        johnsonreg <= #1 4'b00000000;
21      end
22      else begin
23        johnsonreg <= #1 {!johnsonreg[0], johnsonreg[7:1]};
24      end
25    end
26
27    endmodule
```

and test it using the same basic counter test bench created for the simple counter, giving the simulation results as shown in Figure 24.6. We can see that the counter variable "ripples" through till it gets to all 1s and then carries back on until it is all 0s.

24.7 BCD Counter

The BCD (Binary Coded Decimal) counter is simply a counter that resets when the decimal value 10 is reached instead of the normal 15 for a 4-bit binary counter. This counter is often used for decimal displays and other human interface hardware. The VHDL for a BCD counter is very similar to that of a basic binary counter except that the maximum value is 10 (hexadecimal A) instead of 15 (hexadecimal F). The VHDL for a simple BCD counter is given in the following listing. The only change is that the counter has an extra check to reset when the value of the count variable is greater than 9 (the counter range is 0 to 9).

```
1    library ieee;
2    use ieee.std_logic_1164.all;
3    use ieee.numeric_std.all;
4
5    entity counter is
6      generic (
7          n : integer := 4;
8      port (
9          clk : in std_logic;
10         rst : in std_logic;
11         output : out std_logic_vector((n-1) downto 0)
12     );
13   end;
14
15   architecture simple of counter is
16   begin
17     process(clk, rst)
18       variable count : unsigned((n-1) downto 0);
19     begin
20       if rst = '0' then
21         count := (others => '0');
22       elsif rising_edge(clk) then
23         count := count + 1;
24         if count > 9 then
25               count := 0;
26         else if
27       end if;
28       output <= std_logic_vector(count);
29     end process;
30   end;
```

In a similar manner we can implement a BCD counter in Verilog using the code given here:

```
1    module bcd_counter (
2      clk,                  // clock input
3      rst,                  // reset (active low)
4      counter_output   // counter output
5      );
6
7    input clk;
8    input rst;
9
10   output [3:0] counter_output;
11
12   wire clk;
13   wire rst;
14
15   reg [3:0] counter_output ;
16
17   always @ (posedge clk)
18   begin : count
19     if (rst == 1'b0) begin
20       counter_output <= #1 4'b0000;
```

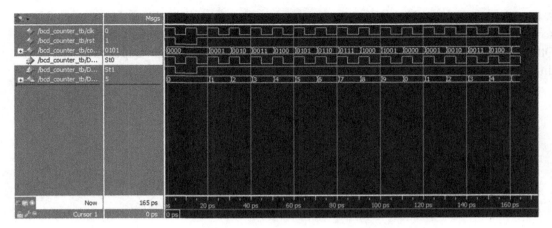

Figure 24.7
BCD counter simulation.

```
21      end
22      else begin
23        if(counter_output < 9) begin
24          counter_output <= #1 counter_output + 1;
25          end
26          else
27            counter_output <= #1 4'b0000;
28      end
29    end
30
31    endmodule
```

and test it using the same basic counter test bench created for the simple counter, giving the simulation results as shown in Figure 24.7. In the results this time you can see the counter variable in binary and also in unsigned decimal counting up to 1001 (binary) and 9 (decimal), then returning back to 0, giving the decimal counter.

24.8 Summary

In this chapter, we have investigated some basic counters and shown how VHDL and Verilog can be used to carry out arithmetic functions or logic functions to obtain the required counting sequence. The possibilities of counters based on these basic types are numerous, possibly infinite, and it is left to the readers to develop their own variations based on these standard types.

A useful exercise would be to modify the basic binary counter by adding an up/down flag so that, depending on this flag, the counter would increment or decrement respectively. Other options would be to extend the shift register to shift left or right depending on a directional flag.

CHAPTER 25

Decoders and Multiplexers

25.1 Decoders

A decoder is a simple combinatorial block that converts one form of digital representation into another. Usually, a decoder takes a smaller representation and converts it into a larger one (the opposite of encoding). Typical examples are the decoding of an n-bit word into 2^n individual logic signals. For example a 3-8 decoder takes three logic signals in and decodes the value of one of the eight output signals (2^3) to the selected value. The symbol for such a decoder is given in Figure 25.1 with its functional behavior shown in the following table:

The VHDL for this decoder uses a simple VHDL construct similar to the *if - else - end if* form, except using the *when - else* syntax. If a signal is assigned a value when a condition is satisfied, then a single assignment can be made using the following basic pseudocode:

```
1    output <= value when condition;
```

This can be extended with else statements to cover a set of different conditions, thus:

```
1    output <=    value1 when condition1 else
2      value2 when condition2 else
3      ...
4      valuen when condition;
```

Finally, if there is a "catch all" condition, similar to the final else in an *if - else - end if* conditional statement in VHDL, then the final assignment would be added as follows:

S2	S1	S0	Q7	Q6	Q5	Q4	Q3	Q2	Q1	Q0
0	0	0	0	0	0	0	0	0	0	1
0	0	1	0	0	0	0	0	0	1	0
0	1	0	0	0	0	0	0	1	0	0
0	1	1	0	0	0	0	1	0	0	0
1	0	0	0	0	0	1	0	0	0	0
1	0	1	0	0	1	0	0	0	0	0
1	1	0	0	1	0	0	0	0	0	0
1	1	1	1	0	0	0	0	0	0	0

Design Recipes for FPGAs. http://dx.doi.org/10.1016/B978-0-08-097129-2.00025-8

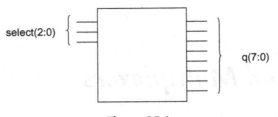

select(2:0)

q(7:0)

Figure 25.1
3-8 decoder.

```
1    output <=   value1 when condition1 else
2      value2 when condition2 else
3      ...
4      valuen when conditionn else
5      valuedefault;
```

Using this approach, the 3-8 decoder can be simply implemented using the following VHDL:

```
1    library ieee;
2    use ieee.std_logic_1164.all;
3    use ieee.numeric_std.all;
4
5    entity decoder38 is
6      port (
7        s : in std_logic_vector (2 downto 0);
8        q : out std_logic_vector(7 downto 0)
9        );
10   end;
11
12   architecture simple of decoder38 is
13   begin
14     q <= "00000001" when s = "000" else
15       "00000010" when s = "001" else
16       "00000100" when s = "010" else
17       "00001000" when s = "011" else
18       "00010000" when s = "100" else
19       "00100000" when s = "101" else
20       "01000000" when s = "110" else
21       "10000000" when s = "111" else
22       "XXXXXXXX";
23   end;
```

The test bench for this decoder could be a simple look-up table of values, but in fact we could combine the clock and reset test bench from the counter example, and include a simple counter in the test bench to generate the signals input to the decoder as follows:

```
1    library ieee;
2    use ieee.std_logic_1164.all;
3    use ieee.numeric_std.all;
4
```

```
5    entity Decoder38Test is
6    end Decoder38Test;
7
8    architecture stimulus of Decoder38Test is
9      signal rst : std_logic := '0';
10     signal clk : std_logic:='0';
11     signal s : std_logic_vector(2 downto 0);
12     signal q : std_logic_vector(7 downto 0);
13
14     component decoder38
15       port(
16         s : in std_logic_vector(2 downto 0);
17         q : out std_logic_vector(7 downto 0)
18       );
19     end component;
20     for all : decoder38 use entity work.decoder38 ;
21
22   begin
23
24     CUT: decoder38 port map(s => s, q => q);
25     clk <= not clk after 1 us;
26     process
27       begin
28         rst<='0','1' after 2.5 us;
29         wait;
30     end process;
31
32     process(clk, rst)
33       variable count : unsigned(2 downto 0);
34     begin
35       if rst = '0' then
36         count := (others => '0');
37       elsif rising_edge(clk) then
38         count := count + 1;
39       end if;
40       s <= std_logic_vector(count);
41     end process;
42
43   end;
```

In Verilog, we can use similar techniques to select individual lines from a binary choice, except that we use a slightly different conditional assignment syntax using a *case* statement:

```
1    case (s)
2      3'h0: q = 8'b00000001;
3      ...
4      default: q = 8'b00000000;
5    endcase
```

The resulting Verilog code is given as follows:

```
1    module decoder38(s, q);
2      output reg [7:0] q;
3      input [2:0] s;
4
5      always @(CharIn)
```

```
 6          case (CharIn)
 7            3'h0: HexOut = 8'b00000001;
 8            3'h1: HexOut = 8'b00000010;
 9            3'h2: HexOut = 8'b00000100;
10            3'h3: HexOut = 8'b00001000;
11            3'h4: HexOut = 8'b00010000;
12            3'h5: HexOut = 8'b00100000;
13            3'h6: HexOut = 8'b01000000;
14            3'h7: HexOut = 8'b10000000;
15            default: HexOut = 8'b00000000;
16          endcase
17        endmodule
```

In Verilog we can use a simple counter in a test bench to select each possibility in turn and then observe the output:

```
 1        module decoder38_tb (
 2          clk,
 3          qout
 4          );
 5
 6        input clk;
 7        output [7:0] qout;
 8
 9        reg [2:0] s = 3'h0;
10
11        always @ (posedge clk)
12        begin
13          // Increment the counter
14          s <= s + 1;
15        end
16
17        // Decode the character into the LED segments
18        decoder38 decoder381(qout, s);
19
20        endmodule
```

25.2 Multiplexers

A multiplexer (MUX) is an extension of a simple decoder in that a series of inputs is decoded to provide select enables for one of a number of inputs. In a similar way that n-bits can decode 2^n signals, in a multiplexer, n bits of select line are required to multiplex 2^n signals. Multiplexers are essential in FPGA internal architectures to select between different implementations of combinatorial logic blocks. For example, consider the simplest multiplexer, a two input (A and B), single output (Q) multiplexer, with a single select line (S). The IEEE symbol for such a MUX is given in Figure 25.2.

A similar approach to the decoder using the *when - else* structure can be used to create a simple implementation of the multiplexer, as shown in the following VHDL:

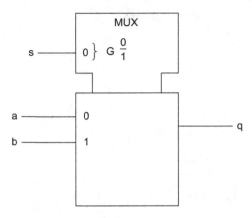

Figure 25.2
Input multiplexer with a single select line.

```
1    library ieee;
2    use ieee.std_logic_1164.all;
3    use ieee.numeric_std.all;
4
5    entity mux21 is
6      port (
7        s : in std_logic;
8        a : in std_logic;
9        b : in std_logic;
10       q : out std_logic
11      );
12   end;
13
14   architecture simple of mux21 is
15   begin
16     q <= a when s = '0' else
17          b when s = '1' else
18          'X';
19   end;
```

This is an extremely useful model and is extensively used in test structures where it is required to choose between a functional and test input signal input to a flip-flop. The model can be easily extended to accommodate multiple input signals. For example, consider a four input multiplexer, with two select signals (inputs = 2select) and a single output. The VHDL model has largely the same structure, but would look like this:

```
1    library ieee;
2    use ieee.std_logic_1164.all;
3    use ieee.numeric_std.all;
4
5    entity mux41 is
6      port (
```

```
7        s : in std_logic_vector (1 downto 0);
8        a : in std_logic;
9        b : in std_logic;
10       c : in std_logic;
11       d : in std_logic;
12       q : out std_logic
13       );
14    end;
15
16    architecture simple of mux41 is
17    begin
18      q <= a when s = "00" else
19           b when s = "01" else
20           c when s = "10" else
21           d when s = "11" else
22           'X';
23    end;
```

Verilog can be used to implement a very similar model, using the select line (s) to define which input (a or b) will be used to set the output (q). The resulting model is shown in the following listing:

```
1     module mux21(s, a, b, q);
2       output q;
3       reg q;
4       input s;
5       input a;
6       input b;
7
8       always @(s or a or b)
9       begin
10        if ( s == 0 )
11          q = a;
12        else
13          q = b;
14        end if
15      end
16    endmodule
```

A more elegant way to accomplish the same function is to declare the input choice using an array (d[2]) as shown in the listing following. This is also an extremely scalable way to implement the function, as the size of the input could be defined by a parameter.

```
1     module mux21b(s, d, q);
2       output q;
3       reg q;
4       input [1:0] d;
5       input s;
6
7       always @(s or d)
8         q = d[s]
9     endmodule
```

25.3 Summary

This chapter has described the basic mechanism for decoding and multiplexing signals using VHDL and Verilog. This is an extremely useful function as it is central to much of the data and control signal management required in complex designs on FPGAs.

CHAPTER 26

Multiplication

26.1 Introduction

A key function in any hardware design that requires signal processing is multiplication. In order to implement such a function it is useful to introduce the basic methods for binary multiplication from first principles so that the implemented approaches can be understood. In this chapter, we will describe these methods and illustrate them with VHDL and Verilog.

26.2 Basic Binary Multiplication

The simplest approach to binary multiplication is essentially long multiplication applied to binary numbers. Consider a basic example of a decimal long multiplication first to remind us of the basic concept, take a multiplication of two numbers 23 and 17:

$$
\begin{array}{r}
23 \\
\times\ 17 \\
\hline
161 \\
023 \\
\hline
0391 \\
\end{array}
$$

This can be implemented using binary numbers in exactly the same way, except instead of decimal numbers, the arithmetic is binary. Consider the multiplication of two unsigned binary numbers for 6 (0110) and 4 (0100). Simply take each bit of the multiplier (4 in this case) and if it is zero, add nothing, but if the bit is one, add the shifted multiplicand (6 in this case). This is illustrated in the binary multiplication below:

$$
\begin{array}{rr}
0110 & 6 \\
0100 & 4 \\
\hline
0000 \\
0000 \\
0110 \\
0000 \\
\hline
0011000 & 24 \\
\end{array}
$$

Design Recipes for FPGAs. http://dx.doi.org/10.1016/B978-0-08-097129-2.00026-X

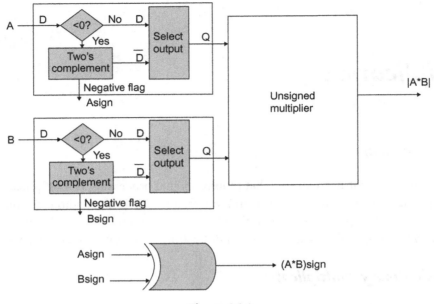

Figure 26.1
Basic signed multiplication.

The way we can often implement this in practice is to have a "partial product" and then add the shifted multiplicand (or zeros) at each stage of the process until the multiplication is complete. While this approach works for unsigned binary numbers, it does not work for two's complement numbers. In the case of two's complement, using a similar approach requires the addition of sign bits to the left of the shifted multiplicand at each stage and then a final step of negating the multiplicand and adding the final shifted value to the partial product. A simpler approach that lends itself well to hardware implementation is simply to test whether a number (or both) are negative, invert to obtain the magnitude of each number if necessary, carry out an unsigned multiplication, then, depending on how many of the arguments are negative, invert the output (two's complement). The method of checking for negative numbers is relatively simple, as an XOR function on the MSB of the two input signed numbers will tag whether the output needs to have a two's complement taken. This is shown schematically in Figure 26.1.

26.3 VHDL Unsigned Multiplier

If we start with a simple unsigned multiplier, then this can be implemented very simply using VHDL. The important aspect to consider with this multiplier is how many bits will be on the inputs and how many on the output. If the number of bits are the same across all three, then we need to consider the possibility of overflow and how this can be handled by the multiplier.

In this basic model, we will define the output number of bits as being the sum of the two input word lengths, and deal with overflow externally to the multiplier.

If we use the basic accumulator and addition function of the simple binary addition method described previously, we can implement a basic VHDL multiplier as shown below:

```
1    library ieee;
2    use IEEE.std_logic_1164.all;
3
4    entity mult_beh is
5      generic(top : natural := 15);
6      port (
7        clk : in std_logic;
8        nrst : in std_logic;
9        a   : in std_logic_vector (top downto 0);
10       b    : in std_logic_vector (top downto 0);
11       product : out std_logic_vector (2*top+1 downto 0)
12     );
13   end entity mult_beh;
14
15   architecture behavior of mult_beh is
16     component add_beh
17     generic (
18       top : integer := 7
19     );
20     port (
21       signal a : in std_logic_vector(top downto 0);
22       signal b : in std_logic_vector(top downto 0);
23       signal cin : in std_logic;
24       signal cout : out std_logic;
25       signal sum : out std_logic_vector(top downto 0)
26     );
27     end component;
28     for all : add_beh use entity work.add_beh;
29
30     signal cin : std_logic := '0';
31     signal cout : std_logic := '0';
32     signal acc : std_logic_vector(2*top+1 downto 0);
33     signal sum : std_logic_vector(2*top+1 downto 0);
34     signal mand : std_logic_vector(2*top+1 downto 0);
35     signal index : integer := 0;
36     signal finished : std_logic := '0';
37   begin
38
39     DUT :add_beh generic map (2*top+1) port map (acc,mand,cin,cout,sum);
40
41     p1 : process (clk, nrst)
42       variable mandvar : std_logic_vector(2*top+1 downto 0);
43     begin
44       if (nrst = '0') then
45         acc <= (others => '0');
46         index <= 0;
47         finished <= '0';
48       else
```

```
49            if clk'event then
50             if clk = '1' then
51              if index <= top then
52                index <= index + 1;
53                mandvar := (others => '0');
54                if b(index) = '1' then
55                  for i in 0 to top loop
56                     mandvar(i+index) := a(i);
57                  end loop;
58                end if;
59              end if;
60              mand <= mandvar;
61              acc <= sum;
62            else
63              if index > top-1 then
64                finished <= '1';
65              end if;
66            end if;
67          end if;
68        end process p1;
69        p2 : process ( finished)
70        begin
71          if rising_edge(finished) then
72            product <= sum;
73          end if;
74        end process p2;
75      end architecture behavior;
```

This model is perhaps more complex than it really needs to be, but it does have some nice features from a learning point of view.

Firstly, rather than a "super efficient" shifting model which is difficult to read, the shift and add function in process *p1* is laid out in detail so each stage of the multiplication can be followed through. Also, notice the use of the signal *finished* which is used to show when the calculation is completed. This is useful when designing a controller to show that the calculation has been completed.

26.4 Synthesis of the Multiplication Function

After completion, this model was run through a standard synthesis software tool, targeted at a smallish size FPGA with the following results:

```
Number of ports :                     66
Number of nets :                     1704
Number of instances :                1639
Number of references to this view :     0
Total accumulated area :
Number of BUFGP :                       1
Number of Dffs or Latches :           164
Number of Function Generators :      1181
Number of IBUF :                       33
```

```
     Number of MUX CARRYs :                 31
     Number of MUXF5 :                      221
     Number of MUXF6 :                      2
     Number of OBUF :                       32
     Number of accumulated instances :      1701
   Number of global buffers used: 1
   ************************************************
   Device Utilization for 2VP2fg256
   ************************************************
   Resource            Used    Avail   Utilization
   ------------------------------------------------
   IOs                 65      140     46.43%
   Global Buffers      1       16       6.25%
   Function Generators 1181    2816    41.94%
   CLB Slices          591     1408    41.97%
   Dffs or Latches     164     3236     5.07%
   Block RAMs          0       12       0.00%
   Block Multipliers   0       12       0.00%
   ------------------------------------------------
   Clock            : Frequency
   ------------------------------------------------


   clk              : 30.0 MHz
   finished         : 30.0 MHz
```

What is clear from this report is the fact that a significant amount of resources was required to implement this multiplier on a small device. In this case, the optimization was for area and not speed, but in spite of that, the design usage was nearly 50% of the whole FPGA. Clearly, arithmetic functions are not always easy on an FPGA, certainly not in area terms, with the worst culprit being multipliers. However, we are going to investigate alternative techniques and this is really just for a comparison of resources.

As a result, care must be taken in managing designs, taking advantage of pipelining and using the available resources as effectively as possible. The downside is that the design becomes more involved, with a controller generally required, but ultimately with the possibility of higher performance than an equivalent DSP function. It is also the case that many modern FPGAs now contain dedicated DSP functions (such as multiplication) which can be targeted directly in synthesis, which means that the area issue will not occur.

26.5 Simple Multiplication using VHDL

As we have seen in the previous example, there is a method of implementing multiplication operations using a "first principles" approach and it is incredibly hungry in terms of both resources and time (taking n shifts to complete a multiplication would lead to a really slow device).

There is, however, an alternative approach with many modern FPGAs that include multiplier blocks as part of the design. These are custom multiplication blocks already in place on the FPGA and this allows the specific multiply function to be implemented directly in the VHDL.

We can therefore convert the std_logic_vector signals into signed signals and then apply the product equation directly using the following VHDL (remember a and b are the two inputs, both of type std_logic_vector, and product is the output, also of type std_logic_vector).

```
1    Product <= std_logic_vector( signed(a) * signed(b) );
```

Clearly this is much more efficient VHDL than the previous model, but also remember that it is necessary to declare the IEEE numeric standard library:

```
1    Use ieee.numeric_std.all;
```

This allows the use of the signed variable types. The complete model using this approach is much more compact and is shown below:

```
1    library ieee;
2    use IEEE.std_logic_1164.all;
3    use ieee.numeric_std.all;
4
5    entity mult_sign is
6      generic(top : natural := 15);
7      port (
8        clk : in std_logic;
9        nrst : in std_logic;
10       a    : in std_logic_vector (top downto 0);
11       b    : in std_logic_vector (top downto 0);
12       product : out std_logic_vector (2*top+1 downto 0)
13       );
14   end entity mult_sign;
15
16   architecture behavior of mult_sign is
17   begin
18     p1 : process (a,b)
19     begin
20       product <= std_logic_vector(signed(a) * signed(b));
21     end process p1;
22   end architecture behavior;
```

The resulting synthesis output is much more compact. Clearly the number of IO blocks (IOBs) will remain the same, but the usage internally on the FPGA will be much reduced:

```
Number of ports :                      66
 Number of nets :                     128
 Number of instances :                 65
 Number of references to this view :    0
Total accumulated area :
 Number of Block Multipliers :          1
 Number of gates :                      0
 Number of accumulated instances :     65
```

```
Number of global buffers used: 0
**************************************************
Device Utilization for 2VP2fg256
**************************************************
Resource              Used    Avail   Utilization
--------------------------------------------------
IOs                    66      140      47.14%
Global Buffers          0       16       0.00%
Function Generators     0     2816       0.00%
CLB Slices              0     1408       0.00%
Dffs or Latches         0     3236       0.00%
Block RAMs              0       12       0.00%
Block Multipliers       1       12       8.33%
```

Clearly, for this device, there are 12 multipliers available, and we have used only one, so the utilization of the remainder of the device is zero. This does lead to the ability to implement certain lower order filters very effectively using devices such as these.

26.6 Simple Multiplication using Verilog

We can use the same simple approach in Verilog as we have just seen in VHDL, where the use of the basic unsigned types and a multiplication can be defined in the code directly, and then the synthesis software will take care of the translation into a physical multiplier.

In the multiplier model in Verilog, we can use a simple assignment as shown in the code snippet following, where the *always* statement is used to check for changes in a or b before assigning the output q to the product. This requires the definition of the output as a register (but we could also use the approach of defining the output as a wire and using the *assign* statement instead).

```
1    always @ (a or b)
2    begin
3      q <= a * b;
4    end
```

The complete multiplier model is given in the following listing:

```
1    module signmult (
2      q, // Multiplication Output
3      a, // Number a
4      b  // Number b
5
6    );
7
8    input [3:0] a; // 4 bit input
9    input [3:0] b; // 4 bit input
10   output [7:0] q; // 8 bit output
11   reg [7:0] q;
12
13   always @ (a or b)
```

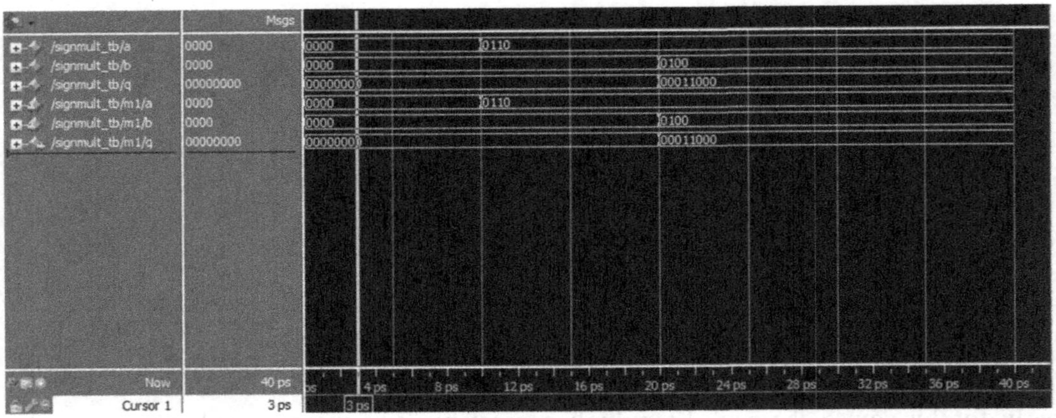

Figure 26.2
Unsigned multiplication of Verilog.

```
14      begin
15        q <= a * b;
16      end
17
18      endmodule
```

In order to test the model we can create a very simple test bench that defines the two variables (a and b), and after initializing them to zero, sets them to 6 and 4, respectively. The resulting simulation results are shown in Figure 26.2, which shows the output start at 0 (0 × 0), remain at 0 (0 × 4) and then finally change to 24 (6 × 4).

```
1       module signmult_tb();
2
3       reg [3:0] a,b;
4       wire [7:0] q;
5
6       signmult m1(q,a,b);
7
8       initial
9         begin
10
11        a = 4'b0000;
12        b = 4'b0000;
13        # 10 a = 4'b0110;
14        # 10 b = 4'b0100;
15
16        $display("a=%d b=%d q=%d\n", a, b, q);
17        end
18
19      endmodule
```

26.7 Summary

This chapter has introduced some techniques for implementing multiplication in VHDL and Verilog for FPGAs and has highlighted the clear difference between using a "first principles" approach as opposed to utilizing the available resources on the FPGA, both in terms of area usage and also in terms of model complexity.

There are, of course, other topologies of multiplier, including the Booth multiplier to name but one, and these are commonly used in hardware. The reader is encouraged to investigate different options for implementing hardware such as multipliers and how best to implement the function for their own application.

CHAPTER 27

Simple 7-Segment (LCD) Displays

27.1 Introduction

Simple 7-segment (LCD) displays are in many respects a simple form of decoder, where an input code of 4 bits is used to specify one of 16 different hexadecimal outputs to be displayed on an LCD module. The LCD module itself usually consists of 7 individual LEDs or lights (hence the alternative name "7-segment display"), which are driven to deliver a particular character as shown in Figure 27.1. Obviously the display can show any combination of the 7 individual bits, with 2^7 combinations; however, in practice they are mostly used to display alphanumeric characters 0-9 and A-F.

In addition to the basic character display, there is also usually a decimal point light, and this has its own control. Therefore, the LCD can be controlled using 8 individual bits; however, to make the designer's job easier, a decoder which uses the correct hexadecimal character as the input (e.g., 0000 for the character *0*) is implemented. This also has the benefit that, for counters, the input character can be determined using arithmetic and then the decoder will handle the translation to the individual segments to be shown. This is a typical *design reuse* task, where once this has been done once, then the designer can reuse the code on any application that uses this type of display.

Finally, depending on the type of module or drive circuit, the logic for the LCD segment being on may be high or low. This will vary from device to device, and in the case of this chapter, the particular development kit being used defined the logic levels as being Low for ON and High for OFF. A typical LCD display has anode and cathode connections to each light emitting diode (LED) in the display and depending on the type of device will have either common cathodes or anodes. This information will be found on the device or development kit data sheet. Using this approach, the coding for a hexadecimal character input to the LCD output is defined in the following table:

27.2 VHDL LCD Module Decoder

The VHDL for a simple LCD decoder uses a simple VHDL construct setting the output depending on a specific condition. If a signal is assigned a value when a condition is satisfied, then a single assignment can be made using the following basic pseudocode:

Design Recipes for FPGAs. http://dx.doi.org/10.1016/B978-0-08-097129-2.00027-1

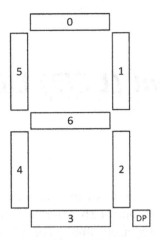

Figure 27.1
7 segment LCD display.

H3	H2	H1	H0	LCD6	LCD5	LCD4	LCD3	LCD2	LCD1	LCD0
0	0	0	0	1	0	0	0	0	0	0
0	0	0	1	1	1	1	1	0	0	1
0	0	1	0	0	1	0	0	1	0	0
0	0	1	1	0	1	1	0	0	0	0
0	1	0	0	0	0	1	1	0	0	1
0	1	0	1	0	0	1	0	0	1	0
0	1	1	0	1	0	0	0	0	1	0
0	1	1	1	1	1	1	1	0	0	0
1	0	0	0	0	0	0	0	0	0	0
1	0	0	1	0	0	1	1	0	0	0
1	0	1	0	0	0	0	1	0	0	0
1	0	1	1	0	0	0	0	0	1	1
1	1	0	0	1	0	0	0	1	1	0
1	1	0	1	0	1	0	0	0	0	1
1	1	1	0	0	0	0	0	1	1	0
1	1	1	1	0	0	0	1	1	1	0

```
1    output <= value when condition;
```

This can be extended with else statements to cover a set of different conditions, thus:

```
1    output <=    value1 when condition1 else
2                 value2 when condition2 else
3                      . . .
4                 valuen when condition;
```

Finally, if there is a "catch all" default condition, then the final assignment would be added as follows:

```
1    output <=    value1 when condition1 else
2                 value2 when condition2 else
3                 ...
4                 valuen when conditionn else
5                 valuedefault;
```

This could also be implemented as a dedicated VHDL function which returned the correct combination of bits. Using this approach, the LCD decoder can be simply implemented using the following VHDL:

```
1    library ieee;
2    use ieee.std_logic_1164.all;
3    use ieee.numeric_std.all;
4
5    entity hexdecoder is
6      port (
7        charin : in std_logic_vector (3 downto 0);
8        hexout : out std_logic_vector(6 downto 0)
9      );
10   end;
11
12   architecture simple of hexdecoder is
13   begin
14    hexout <= "1000000" when charin = "0000" else
15      "1111001" when charin = "0001" else
16      "0100100" when charin = "0010" else
17      "0110000" when charin = "0011" else
18      "0011001" when charin = "0100" else
19      "0010010" when charin = "0101" else
20      "0000010" when charin = "0110" else
21      "1111000" when charin = "0111" else
22      "0000000" when charin = "1000" else
23      "0011000" when charin = "1001" else
24      "0001000" when charin = "1010" else
25      "0000011" when charin = "1011" else
26      "1000110" when charin = "1100" else
27      "0100001" when charin = "1101" else
28      "0000110" when charin = "1110" else
29      "0001110" when charin = "1111" else
30      "0110110";
31   end;
```

The test bench for this decoder could be a simple look-up table of values, but in fact we could combine the clock and reset test bench from the counter example, and include a simple counter in the test bench to generate the signals input to the decoder as follows:

```
1    library ieee;
2    use ieee.std_logic_1164.all;
3    use ieee.numeric_std.all;
4
5    entity test1 is
6      port (
7        rst : in std_logic;
8        clk : in std_logic;
9        hexout : out std_logic_vector(6 downto 0)
```

```
10      );
11    end test1;
12
13    architecture stimulus of test1 is
14      signal charin : std_logic_vector(3 downto 0);
15
16      component hexdecoder
17        port(
18          charin : in std_logic_vector(3 downto 0);
19          hexout : out std_logic_vector(6 downto 0)
20        );
21      end component;
22      for all : hexdecoder use entity work.hexdecoder ;
23
24    begin
25
26      CUT: hexdecoder port map(charin => charin, hexout => hexout);
27
28      process(clk, rst)
29        variable count : unsigned(26 downto 0);
30          variable charcount : unsigned(3 downto 0);
31      begin
32        if rst = '0' then
33          count := (others => '0');
34          charcount := (others => '0');
35        elsif rising_edge(clk) then
36          if count = 50000000 then
37            count := (others => '0');
38            charcount := charcount + 1;
39          else
40            count := count + 1;
41          end if;
42        end if;
43        charin <= std_logic_vector(charcount);
44      end process;
45    end;
```

27.3 *Verilog LCD Module Decoder*

In Verilog the approach is very similar to the VHDL, except that we use a slightly different conditional assignment syntax using a *case* statement:

```
1    case (CharIn)
2      4'h0: HexOut = 7'b1000000;
3      ...
4      default: HexOut = 7'b0110110;
5    endcase
```

The resulting Verilog code is given here:

```
1    module HexDecoder(HexOut, CharIn);
2      output reg [6:0] HexOut;
3      input [3:0] CharIn;
4
5      always @(CharIn)
```

```
6        case (CharIn)
7          4'h0: HexOut = 7'b1000000;
8          4'h1: HexOut = 7'b1111001;
9          4'h2: HexOut = 7'b0100100;
10         4'h3: HexOut = 7'b0110000;
11         4'h4: HexOut = 7'b0011001;
12         4'h5: HexOut = 7'b0010010;
13         4'h6: HexOut = 7'b0000010;
14         4'h7: HexOut = 7'b1111000;
15         4'h8: HexOut = 7'b0000000;
16         4'h9: HexOut = 7'b0011000;
17         4'hA: HexOut = 7'b0001000;
18         4'hB: HexOut = 7'b0000011;
19         4'hC: HexOut = 7'b1000110;
20         4'hD: HexOut = 7'b0100001;
21         4'hE: HexOut = 7'b0000110;
22         4'hF: HexOut = 7'b0001110;
23         default: HexOut = 7'b0110110;
24       endcase
25     endmodule
```

And, again this is tested using a simple counter to display each character in sequence. In this example, the FPGA is running with a clock frequency of 50 MHz, so by setting a counter to 50×10^6 the increment will take place at roughly one-second intervals and display each character in turn. The complete test bench is shown in this listing:

```
1      module test1 (
2        clk,
3        hex0,
4        dp0
5        );
6
7        // Configuration parameters
8      localparam HB_CNT_WIDTH = 26;
9      localparam HB_CNT_MSb = HB_CNT_WIDTH - 1;
10
11     input clk;
12     output [6:0] hex0;
13     output dp0;
14
15     reg [ HB_CNT_MSb:0] hbled_r = {HB_CNT_WIDTH{1'b0}};
16     reg [3:0] CharIn = 4'h0;
17     reg dp0;
18
19     always @ (posedge clk)
20     begin
21       // Set the decimal point High which is OFF for this module
22       dp0 = 1'b1;
23
24
25
26       // When the counter reaches zero - increment the character shown
27       if (hbled_r == 50000000) begin
28         CharIn <= CharIn + 4'b0001;
29             hbled_r <= 0;
```

```
30      end else begin
31        // Increment the counter
32        hbled_r <= hbled_r + 1'b1;
33      end
34    end
35
36    // Decode the character into the LED segments
37    HexDecoder HexDecoder1(hex0, CharIn);
38
39    endmodule
```

27.4 Summary

This short chapter has described the basic mechanism for driving simple 7-segment displays using VHDL and Verilog. This is an extremely useful function when using development boards, as it allows a visual representation of data to the user in real time, which can be very helpful for debugging or providing information to the user while the FPGA is running, or status codes to indicate states or particular behavior.

Bibliography

Introduction

It is normal in a book such as this to have a bibliography that simply lists a series of books. However, in this book I have decided to not only list the book titles and details, but also give my perspective on their applicability and context to help the reader in deciding which would be a suitable book for them. Of course, this is limited to my own viewpoint and others may well disagree with my short synopses of the books, but hopefully it will help the reader understand where I found each book useful in this work. There is also a more typical "academic style" bibliography including not only these texts, but any research material that I have found useful in the writing of this book. There are numerous other texts available and readers can find numerous examples online now, many in e-book form, making it easy to obtain the information they require.

Useful Texts for VHDL

Digital Systems Design with VHDL

Digital system design with VHDL by Zwolinski [1], published by Pearson Education, is a superb introduction to designing with VHDL. It is used in many universities worldwide for teaching VHDL at an undergraduate level and has numerous basic examples to enable a student to get started. I would also recommend this to an FPGA engineer getting started with VHDL.

The Designer's Guide to VHDL

The Designer's Guide to VHDL by Ashenden [3] is perhaps the most comprehensive book on VHDL from a variety of perspectives. It covers the syntax and language rigorously, has plenty of examples, and is a great desktop reference book. For nonbeginners in VHDL, this is the book I would recommend.

Design Recipes for FPGAs. http://dx.doi.org/10.1016/B978-0-08-097129-2.10000-5

VHDL: Analysis and Modeling of Digital Systems

VHDL: Analysis and Modeling of Digital Systems (McGraw-Hill Series in Electrical and Computer Engineering) by Navabi [2] is a detailed look at not only how VHDL can be used to model digital systems, but many of the detailed issues regarding timing and analysis that are often skipped over by other texts on VHDL. It is perhaps not a beginner's book, but is especially useful for those who require a deeper understanding of issues relating to timing.

VHDL for Logic Synthesis

VHDL for Logic Synthesis by Andrew Rushton, published by Wiley, Rushton [6] is a useful background text for those who perhaps need to understand how VHDL can be used for practical synthesis. The book discusses what is and what is not synthesizable and also explains how some useful and somewhat arcane VHDL functions operate.

Useful Texts for Verilog

Digital Systems Design with SystemVerilog

Digital System Design with SystemVerilog by Zwolinski [4], published by Pearson Education, is a superb introduction to designing with SystemVerilog. It is used in many universities worldwide for teaching SystemVerilog at an undergraduate level and has numerous basic examples to enable a student to get started. I would also recommend this to an FPGA engineer getting started with SystemVerilog.

Verilog Designer's Library

Verilog Designer's Library by Zeidman [5] is a really useful textbook dedicated to Verilog building blocks for use in real designs. It is mainly targeted at IC design, but of course the main techniques will generally apply just as well to FPGA design.

Useful Texts for FPGAs

The Design Warriors Guide to FPGAs

The Design Warriors Guide to FPGAs by Clive "Max" Maxfield, published by Elsevier, Maxfield [7] is an excellent introduction to the field of FPGAs. It introduces the main concepts in designing with FPGAs as the platform and does not get into low-level details of VHDL or Verilog, but does have a balance between high-level design issues and low-level details. This is especially useful for the student who needs to know how FPGAs work and also for engineers who need a "heads up" on how FPGAs can be used in practice.

General Digital Design Books

Digital Design

Digital Design by M. Morris Mano, published by Prentice Hall, Mano and Ciletti [8] is a good background text for digital design and computer design. A particularly useful aspect for those designing embedded processors is the section of the book that discusses the difference between high-level languages, assembly language and machine code and then develops that into a design methodology. For anyone starting out with processor design, this is a very useful text. Mano also has a related book called *Computer System Architecture* that has more detail in this area and is equally useful.

References

[1] M. Zwolinski, Digital Systems Design with VHDL, second ed., Pearson Education, England, 2003, ISBN 0-13-039985-X.
[2] Z. Navabi, VHDL: Analysis and Modeling of Digital Systems, McGraw-Hill, New York, NY, 1992, ISBN 978-0070464728.
[3] P. Ashenden, Designers Guide to VHDL, Morgan Kaufmann Publishers, San Francisco, CA, 1995, ISBN 1-55860-270-4.
[4] M. Zwolinski, Digital Systems Design with SystemVerilog, Pearson Education, England, 2009, ISBN 978-0137045792.
[5] B. Zeidman, Verilog Designer's Library, Prentice Hall, Upper Saddle River, NJ, 1999, ISBN 0-13-081154-8.
[6] A. Rushton, VHDL for Logic Synthesis, Wiley, New York, NY, 2011, ISBN 978-0470688472.
[7] C. Maxfield, The Design Warrior's Guide to FPGAs, Newnes, Burlington, MA, 2004, ISBN 978-0-7506-7604-5.
[8] M.M. Mano, M. Ciletti, Digital Design, Prentice Hall, Upper Saddle River, NJ, 2006, ISBN 978-0131989245.
[9] P.J.M. Laarhoven, E.H.L. Aarts, Simulated Annealing: Theory and Applications, Kluwer Academic Publishers, Dordrecht, 1989.

Further Reading

A.D. Brown, D. Milton, A. Rushton, P.R. Wilson, Behavioural synthesis utilising recursive definitions, IET Comput. Digit. Tech. 6 (6) (2012) 362–369.
M. Sacker, A.D. Brown, A.J. Rushton, P.R. Wilson, A behavioral synthesis system for asynchronous circuits, IEEE Trans. Very Large Scale Integr. Syst. 12 (9) (2004) 978–994.
P.R. Wilson, A.D. Brown, DES in 4 days using behavioural modeling and simulation, in: IEEE International Behavioral Modeling and Simulation Conference, BMAS 2005, San Jose, USA, 2005.

Index

Note: Page numbers followed by *f* indicate figures and *t* indicate tables.

Printed in the United States
By Bookmasters